mercury lamp

iron arc

barium

calcium

fraunhofer lines

tungsten lamp

fluorescent lamp

Courtesy of Bausch & Lomb

chemistry the science and the scene

chemistry

the science and the scene

ronald d. clark and robert l.s. amai

New Mexico Highlands University

HAMILTON PUBLISHING COMPANY
SANTA BARBARA, CALIFORNIA

Library of Congress Cataloging in Publication Data:
Clark, Ronald D.   1938–
　Chemistry: the science and the scene.

　1. Chemistry. 2. Chemistry, Technical. I. Amai,
Robert L. S., 1932– joint author. II. Title.

QD31.2.C57　　540　　74-13782

ISBN 0-471-15857-7

Printed in the United States of America
10 9 8 7 6 5 4 3 2 1

about the authors

Ronald D. Clark was born in Hollywood, California. He received a B.S. degree in chemistry from the University of California, Los Angeles, in 1960 and a Ph.D. degree in organic chemistry from the University of California, Riverside, in 1964. After a year of post-doctoral work at Michigan State University, Dr. Clark joined the Standard Oil Company (Ohio) as a research chemist, where he was a project leader. He joined the faculty at New Mexico Highlands University in 1969 and is now Associate Professor of Chemistry. Among the courses that he has taught at New Mexico Highlands University are chemistry for nonscience majors, organic chemistry, biochemistry, and industrial processes chemistry. His patents and published papers cover such diverse topics as stereochemistry of small rings, catalysis, and industrial research. Dr. Clark is a member of the American Chemical Society and the Chemical Society (London) and has been active in the Central New Mexico Section of the American Chemical Society.

Robert L. S. Amai, Professor of Chemistry at New Mexico Highlands University, is a native of Hilo, Hawaii. He received his B.A. (1954) and M.S. (1956) degrees in chemistry from the University of Hawaii. His Ph.D. degree in organic chemistry was awarded by Iowa State University in 1962. Dr. Amai joined the research staff at New Mexico Highlands University in 1960, and in 1962 he became a member of the faculty of the Department of Chemistry. From 1969 to 1974, he served as department chairman. He has been director of a sequential summer institute in natural science for secondary school science and mathematics teachers, sponsored by the National Science Foundation, a director of the Northeastern New Mexico Regional Science Fair, and a participant in the Visiting Scientist Program of the New Mexico Academy of Science. Among his teaching responsibilities are the chemistry course for nonscience majors and courses in organic chemistry, biochemistry, and the chemistry of medicinal compounds. His research interests include the synthesis of compounds of potential medicinal use and the study of natural plant products. Dr. Amai is a member of the American Chemical Society, the American Association for the Advancement of Science, the Society of the Sigma Xi, and the New Mexico Academy of Science.

preface

Selecting a text for a short course in chemistry for nonmajors is difficult. The type of student who will use the text, the objectives of the course, and the philosophy of the instructor are just a few of the factors affecting the choice. Several years ago we were faced with this problem of selecting a text. At the time, none of the available publications seemed appropriate. Some were too long for our course. Others concentrated too heavily on either the one extreme of chemical principles or the other extreme of all application with insufficient chemical principles to give a firm foundation.

As with most courses of this type, our course was not designed to train chemists. For this reason, the course was not intended to be an abbreviated version of the chemistry majors' course. It was felt that the course should be sufficiently well-founded in chemical principles so that the student would be able to get a feeling for the *science* of chemistry. At the same time, however, it was recognized that there is a tremendous technology of chemistry in the real world and that the student should have an introduction to this aspect of chemistry as well.

It was with these aims in mind, then, that we decided to write our own text: one that would blend all these features into a harmonious unit. We quickly ran into the problems faced by others who have tried to achieve these same goals. The selection of topics to include and the manner in which these topics are treated had to be carefully considered. The extent of integration of practical applications of chemistry with the necessary theoretical materials was also a major problem. Finally, we recognized that student interests change and that therefore a great variety of material must be included.

The first eleven chapters in the text cover basic concepts of chemistry. The emphasis is on atomic and molecular structure and chemical bonding, which we feel are of major importance to the understanding of chemical properties and reactions. Other topics discussed in these chapters include the properties and transformations of matter, the metric system of measurements, oxidation-reduction reactions, and acids and bases. The remaining nine chapters of the text cover selected areas of practical or applied chemistry. Topics in this section are organic chemistry and biochemistry, medicinal chemistry, household chemistry, and plastics. Also included are chapters on air and water pollution, the energy situation, and space chemistry, which are topics of current interest and concern to the public and in which chemistry plays a significant role. Some topics, such as nomenclature, gas laws, and balancing complex redox reactions, traditionally included in beginning chemistry texts, have been omitted in favor of topics we felt would be of greater value and interest to typical students in this course.

By the appropriate selection of chapters, this text can be used for a one-quarter, a one-semester, or a two-quarter course, and for students who want to explore current issues, who prefer a more well-rounded exposure to chemistry, or who need a somewhat more rigorous foundation in chemical principles as preparation for further work in a science-related profession. The various options and suggested schedules of topics are described more fully in the Instructor's Manual that accompanies this text.

No prior training of the student in chemistry is assumed for this text. The mathematical requirement is not high, but students are expected to be familiar with basic manipulations. There are only two chapters in which mathematical calculations constitute a major part—Chapter 2, on systems of measurement, and Chapter 9, on the quantitative aspects of chemistry. The latter chapter may be omitted without major consequence to the topics that follow. The text is written in language that is not high-powered or sophisticated, but at the same time does not talk down to the student. Study aids to the student are provided by ample definitions of new terms and phrases when they are first introduced. In addition, a list of these important terms is provided at the beginning of every chapter to alert the student of their introduction in the chapter and to serve as a review list later on. Questions and problems are provided at the end of each chapter and are divided into two groups: one involving questions based directly on chapter material and requiring little or no math and the other containing problems requiring a greater analysis of chapter material or a better proficiency in math. A teacher may selectively make assignments from one or the other or both sets, depending on the nature of the class being taught.

We hope that this text will be a useful tool in making the student's brief look at chemistry both enjoyable and exciting. We also hope that it will help him see the place of chemistry in both the science and technology of the present and the future.

We have many to thank for their contributions to our effort: the staff of Hamilton Publishing Company for their faith, encouragement, and most generous assistance in all phases of the preparation of the text; our reviewers—Ned A. Daugherty, Richard L. McDonald, Raymond F. O'Connor, Robert L. Taylor, Karen C. Timberlake—who provided us with valuable suggestions and criticisms; Laura Tocco for patiently typing our manuscript material, through several versions and revisions; and our students in Chemistry 100 who allowed us to test our manuscript in class and who provided us with the initial impetus to write this book.

R. D. C.
R. L. S. A.

contents

chapter **6**   **electronic structure of the atom**   99

chapter **7**   **the periodic table**   111

chapter **18**   # chemistry of water quality and solid waste 299

chapter **19**   # chemistry and the energy crisis 319

chapter **20**   # chemistry in space 329

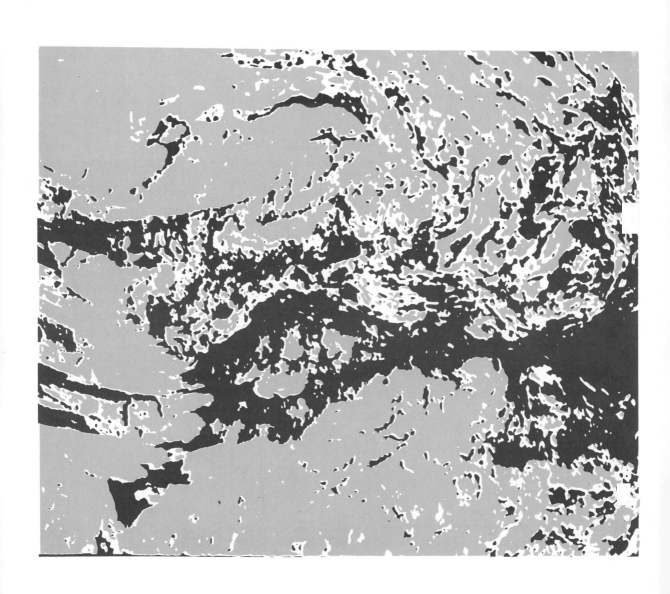

prolog

Over the years, science has received a mixed reception from the general public. Not so long ago, new developments often were viewed with suspicion. Many people resisted the telephone, for example, as an invasion of privacy. As the rate of discoveries increased, however, people's attitude began to change. By the end of World War II, discoveries were being made rapidly and were assimilated easily into society. The American people accustomed to such wonders as the radio, the automobile, the airplane, and the Bomb were able to accept color television and polyester fibers as natural and unsurprising in a country that leads the world in scientific development.

The passive attitude of the American people toward science changed in the late 1950s, when Russia launched the first man-made earth satellite and, with it, the space age. Suddenly, the accustomed superiority of the United States in science was questioned. At the same time, people began to realize that more and better science could solve some of the problems that had plagued mankind for centuries. Weather could be modified to prevent floods and droughts. Fertilizer, pesticides, and new plant varieties could be developed to increase food production. Plastics and other new materials could revolutionize housing. The list is endless. The nation responded with a scientific push that had not been experienced before. Science programs were expanded and then expanded again. Scientifically trained people were not available fast enough to fill the new positions. The confidence placed in science appeared to be confirmed as, step by step, we slowly made our way to the moon.

Meanwhile, people began to have second thoughts about the potential of science as they became aware of the systematic destruction of Vietnam and the deterioration of cities. The environmental crisis finally captured widespread public attention by 1970. As part of the environmental crisis, even such commonplace items as plastic bottles and detergents came under fire. Thus the tide of public opinion about science gradually shifted in the late 1960s to the point that science was seen not as the savior of mankind, but rather as its nemesis. With the advent of the energy crisis in the early 1970s, the attitude of society toward science appeared to moderate somewhat. It is now accepted that technological advancement should be accompanied by increased social responsibility.

Obviously, neither of the extreme extrapolations of science's potential is accurate. Science is not the solution to all man's problems, nor is it the cause of them all. Actually, science should not even be directly involved in the discussion. Science is merely a systematic search for knowledge. Knowledge alone cannot pollute the atmosphere or devastate the cities. The problems often arise when knowledge generated by science is applied thoughtlessly or carelessly by man in an effort to gain benefits for himself or for a special interest group. In other words, it is technology that is responsible for the great benefits man has received from science as well as the great problems that have been created by scientific discoveries. The studies of atomic physics in the first half of the twentieth century added immeasurably to man's understanding of the structure of matter. It was not until World War II, however, that it was thought *desirable* that this new knowledge be applied to the manufacture of the atomic bomb.

The distinction between science and technology is more than semantic. The search for knowledge increases our understanding of the world around us. This new knowledge helps to eliminate superstition and gives us a feeling of security. It makes technological advances possible, but it does not make them mandatory. Furthermore, normally it does not evaluate technological advances. Evaluation must be done by man. To make intelligent decisions, man must be informed of both the advantages and the disadvantages of a technological advance. This means that a technologically oriented society requires a technologically aware population.

In this text we shall present some selected topics in both the science and the technology of chemistry. Enough of the basic science of chemistry is included to permit understanding of the technological basis of many common chemical products and processes. Where appropriate, some of the social implications of technology also are discussed briefly.

chapter **1**

fundamental concepts

When a neon sign is turned on, it produces a characteristic red-orange glow. When a person has acid indigestion, he takes some bicarbonate of soda to relieve it. Soaps form an insoluble scum in hard water, although detergents do not do this as readily. Tableware may be silver-plated or an automobile bumper, chrome-plated. A lead storage battery such as a car battery supplies electricity, but when it is weak it can be recharged by supplying electricity to it. Water can be represented by the formula H_2O, but not by HO or H_2O_2.

All these facts are known to practically everyone and are all seemingly unrelated. There is one thing that they all have in common, however. They all involve chemistry, and each fact can be explained by an appropriate chemical concept or principle. These examples show that chemistry is not something confined to a laboratory, but instead touches on many aspects of everyday living.

In the study of chemistry that is developed in this book, all principles related to the situations given above are included among those that will be discussed. A chemical understanding of these phenomena as well as of other facts and observations about matter that makes up the world then will be possible. For this look at chemistry, it will be satisfactory to define chemistry as the study of the properties and transformations of matter. Although this is a rather short definition, it nevertheless includes concepts that are fundamental to the understanding of all aspects of chemistry. These fundamental concepts are introduced in the initial chapters of this book.

chemistry
scientific method
hypothesis
theory
matter
heat
temperature
kinetic energy
energy
pressure
element
compound
mixture
physical property
chemical property
physical change
chemical change
Law of Conservation of Mass
Law of Conservation of Mass-
 Energy

scope of chemistry

The science of chemistry is not new. As a matter of fact, it has been known since at least the time of ancient civilizations. Of course these early peoples did not have full knowledge of the theories and

Nickel chrome-plating is used extensively as decoration on automobiles, especially for bumpers. (*Courtesy International Nickel Company, Inc.*)

Use of lead storage batteries in cars may be prolonged by recharging the batteries. (*Photograph by A. Marshall Licht*)

concepts of science that are known today. They certainly were aware at least of the gross characteristics and effects of chemical processes and used them to great benefit in their daily lives. Preparation of foodstuff, fabrics, and metal goods, use of fuels and medicines, and processes such as glass production, leather tanning, and wine making are all examples of situations where chemistry and chemical processes were involved even in ancient times.

Chemistry as a field of science is quite broad and complex in its coverage of activities. Five somewhat arbitrary subdivisions of chemistry have been recognized traditionally: analytical, inorganic, organic, physical, and biochemistry. In the past these subdivisions have been useful for organizing the accumulating knowledge in chemistry. Increasingly, however, these subdivisions are proving inadequate. Scientists in all branches of chemistry, for example, employ the tools of analytical chemistry. Physical chemists and even inorganic chemists frequently work with organic molecules. Organic chemists use the principles of physical chemistry to learn about organic chemical reactions.

The designation of chemistry itself as an independent discipline is actually somewhat arbitrary. To truly understand the processes occurring in living cells, biochemists must be well versed in all disciplines of chemistry as well as of biology and physics. Molecular biologists study the chemical aspects of biology, and chemical physicists study the chemical aspects of physics.

the scientific method

When a chemist studies the properties and transformations of matter, he does so by experimental means. Therefore, chemistry often is referred to as an *experimental* science. Every basic concept and law in chemistry has developed from an experimental situation, and no concept or law is accepted unless it has been studied fully and confirmed through rigorous experimental tests. The process by which this is done is called the scientific method. The scientific method is a reasoning or problem-solving process through which a logical and reasonable conclusion is reached. It includes the following basic steps:

1 Making observations and collecting data
2 Developing a hypothesis (a generalization or a tentative idea or conclusion) based on analysis of the data
3 Testing the hypothesis through further experimentation or application

4 Modifying the hypothesis, and further experimentation if necessary

5 Proposing a theory (or explanation) based on the best available information

Although this process is called the scientific method, it is by no means restricted in its use only to scientific situations in laboratories. The process is called *scientific* mainly because it is logical. It is used in making any decision, regardless of whether it is related to a scientific situation or not. For example, in deciding how to spend a Saturday evening, you might first see what movies are showing in town, what concerts or special events are scheduled, how much money you have, or what your friends are planning to do. Based on this information, you then might decide that you would like to see an award-winning play that is being performed downtown. As you pursue this decision further, however, you find that tickets to the play are scarce and the ones that are still available are priced too high for your pocketbook. You therefore need to change your plans, and you make a new decision, based on this additional information, to attend a movie instead. Similar situations about deciding what clothes to wear or what food to serve to dinner guests involve applications of the scientific method to reach the best conclusions.

matter

The "raw material" used in a chemical study is called matter. Matter can be described as "the stuff that the universe is made of," but a more scientific definition would be to call matter anything that has mass or weight and occupies space. A substance need not be heavy, large, or visible to the naked eye to be labeled matter, as long as it meets the two qualifications of having mass and occupying space.

Matter can exist in three states: solid, liquid, or gas. In the solid state, matter particles are closest together and have the least degree of motion. There are relatively strong forces of attraction between particles in a solid, and these are responsible for solids' maintaining definite volumes and shapes.

When matter exists in the liquid state, particles have a greater degree of motion and have the ability to migrate or diffuse throughout the body of the liquid. The attractive forces between particles are not as strong as those that are found in solids. Therefore, although liquids have definite volumes, they do not have definite shapes; they assume the shape of the container in which they are placed.

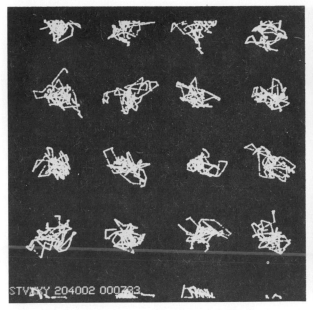 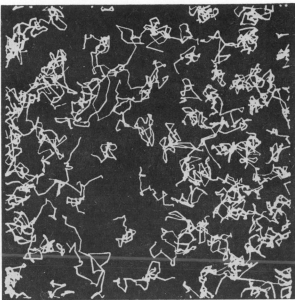

Computer simulations of particle motion displayed on the face of a cathode-ray tube. *Left*: Pattern shows that particles in a solid can move only around well-defined positions. *Right*: Pattern shows that particles in a liquid have a greater degree of motion and are freer to travel from one location to another. (*Courtesy B. J. Alder and T. Wainwright, Lawrence Radiation Laboratory, Livermore, California*)

Particles that have the greatest degree of motion are found in gases. Here the distances between particles are large, and the attractive forces between them are small. Because of the large distances between individual particles, gases are highly compressible and have neither definite volume nor definite shape.

The relative motion of particles in matter is influenced by heat, and the state in which a particular sample of matter exists can be controlled by regulating its temperature. Heat is due to the agitation of particles in a substance, and temperature is a measurement of the intensity of this heat. At lower temperatures, the motion of particles generally is not great, and particles are not so far apart as they are at

---

Heat was once thought to be a form of matter, a fluid, that flowed from hot objects to cooler ones. The name *caloric*, derived from the Latin word *calor* meaning heat, was given to this fluid.

---

higher temperatures. Increasing the temperature increases the kinetic energy of particles, causing them to become more vigorous in their motion. Kinetic energy is the energy of motion, whereas energy itself is the capacity of a system to do work. Increase in motion of particles causes a weakening of the effects of attractive forces that hold the particles to one another and eventually causes a change in state from solid to liquid or from liquid to gas. The increase in kinetic

The element gallium has a relatively low melting point (30°C). Body heat is sufficient to melt a piece of solid gallium, as shown here. This change in state is an example of physical change. (*Courtesy Aluminum Company of America*)

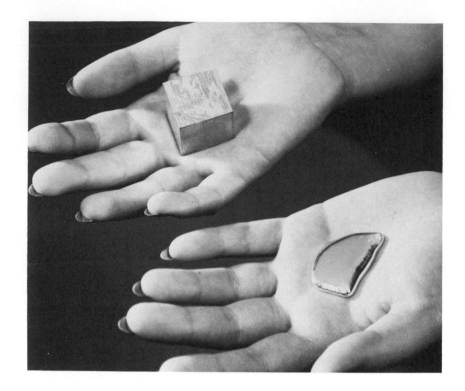

energy also accounts for the observed expansion in the volume of matter as it is heated. These concepts are well illustrated by the characteristics of the three states of water—ice, liquid water, and water vapor or steam—and the changes that occur as ice melts or water boils.

The pressure that is attributed to a gas, particularly in a confined area, is due to the bombardment of moving gas particles on a surface. The more frequent the collisions and the harder the particles hit the surface, the greater is the observed pressure. Consequently, increasing the number of gas particles in a given volume or increasing the temperature of a gas causes an increase in the observed gas pressure.

Matter also may exist in chemically pure and impure forms. Pure matter consists of substances called elements and compounds, while impure matter consists of mixtures. Components of mixtures are themselves elements or compounds in a wide variety of possible combinations. Further discussions of elements, compounds, and mixtures will be found in Chapter 3.

Thus far, the terms *study* and *matter* included in the definition of chemistry have been discussed. The two remaining terms that need discussion are *properties* and *transformation*.

Table salt alone is chemically pure matter, but the mixture of table salt and pepper is impure matter. (*Photographs by A. Marshall Licht*)

properties of matter

Properties of matter fall into two categories—physical and chemical. Physical properties include such characteristics as freezing and boiling points, state, density, size, and shape. Generally these properties refer to the physical existence of matter, rather than to its chemical makeup. External or environmental factors often influence matter's physical properties more than its chemical properties. Changing atmospheric pressure can influence a boiling point, for example, or changing the temperature can influence the state of matter. The measurement or observation of a physical property does not involve changes in the chemical composition of matter, so no new and different substance results in such measurements or observations.

Chemical properties, on the other hand, are related to the internal, chemical makeup of matter. Such characteristics as combustibility, acidity, basicity, and reactivity are based on the manner in which a substance is put together chemically. For example, the fact that sugar and wood can be burned although salt and granite cannot is due to differences in the chemical properties of these substances. The particular components in the makeup of a substance and the structural relationship of these components to one another influence the chemical properties of a substance.

Measurement or observation of chemical properties involves changes in chemical composition; consequently different substances result in the process. One could not observe the chemical reactivity of one substance toward another without actually permitting a chemical reaction to occur between the two substances. This process brings about the conversion of these materials into different substances because of changes in their chemical makeup.

transformations of matter

The changes or transformations that matter can undergo also fall into two categories—physical and chemical. If a transformation occurs in which no change in chemical composition takes place, the transformation is referred to as a physical change. Changes in state, size, shape, degree of subdivision, and so forth are physical changes. Examples of such changes would be melting ice, evaporating water,

Earth-moving equipment creates a physical change in the landscape, but not a chemical change. (*Courtesy Marathon LeTourneau Company, Longview Division*)

crushing rocks, or dissolving sugar in water. In each of these changes, the original substance maintains its chemical identity and characteristics throughout.

If a change of matter creates a different substance, then a change in chemical composition is indicated, and the change is called a chemical change. Burning paper, souring milk, and detonating dynamite are examples of this kind of change. In each situation, new substances result, with properties unlike those of the starting materials.

Many chemical changes are accompanied by physical changes, particularly changes in state or degree of subdivision. When both types of changes occur, the chemical one generally is considered more significant, and the change is referred to as chemical. Thus combustion processes are called chemical changes, even though gaseous products are produced from solid or liquid starting materials.

In the vast majority of transformations of matter, conservation of material is observed. In other words, in ordinary changes, matter is neither created from nonmatter nor destroyed into nonmatter, but it may be changed from one type or form into another. This is a statement of the Law of Conservation of Mass. For example, if 3 ounces of carbon were burned to form carbon dioxide, it would be found that 8

Tearing paper produces only a physical change, but burning paper produces a chemical change. (*Photographs by A. Marshall Licht*)

Combustion, a chemical change, occurs during a forest fire. (*Courtesy U.S. Forest Service*)

In boiling a liquid, heat energy is converted to kinetic energy. A turned-on lamp is an example of energy conversion.
(*Photographs by A. Marshall Licht*)

ounces of oxygen would be needed and 11 ounces of carbon dioxide would be formed.

3 ounces of carbon and 8 ounces of oxygen	FORM	11 ounces of carbon dioxide

Likewise, if 11 ounces of propane gas were burned, 40 ounces of oxygen would be consumed, and 33 ounces of carbon dioxide as well as 18 ounces of water would be formed.

11 ounces of propane and 40 ounces of oxygen	FORM	33 ounces of carbon dioxide and 18 ounces of water

Note that in these examples the total weight of the starting materials used equals the total weight of products formed. The Law of Conservation of Mass applies to all ordinary transformations of matter, regardless of the states of the starting materials or the final products.

Energy also is involved in all ordinary transformations of matter. It is manifested in various forms, such as kinetic energy (energy of motion), potential or stored energy (of which chemical energy is a variety), heat, light, and electrical energy. When energy becomes involved in an ordinary change of matter, it is neither created from nonenergy sources nor destroyed into nonenergy forms. Like matter, it may be transformed from one type into another. When phosphorus

---

The work of Antoine Lavoisier, a French nobleman, probably contributed more than any other study to the establishment of modern chemistry. It was Lavoisier who first showed that oxygen is a constituent of air. He also showed that when sulfur and phosphorus are burned the products actually increase in weight rather than decrease, as was commonly thought at the time. This and other experiments led Lavoisier to recognize the Law of Conservation of Mass. Unfortunately, Lavoisier's work was ended abruptly when he was beheaded on the guillotine in 1794, during the French Reign of Terror!

---

undergoes spontaneous combustion in air, chemical or potential energy is released as heat and light; when a liquid is boiled, heat energy is converted to kinetic energy. A law of conservation similar to that for matter exists for the energy components in matter transformations.

The only deviations from the laws of conservation of both mass and energy are found in radioactive processes and nuclear reactions. In these processes, there is some conversion of a small amount of matter into energy rather than into another form of matter. In some respects, in these processes, matter is destroyed and energy is created in what appear to be violations of the laws of conservation. A more general law incorporates this apparent discrepancy: the Law of Conservation of Mass-Energy. According to this law, mass and energy are interconvertible. Einstein's famous equation, $E = mc^2$, quantitatively relates energy (E) to mass (m) and the velocity of light (c). Further discussion of this concept appears in Chapter 5.

In order that scientific findings can be accurately interpreted, recorded, and transmitted, it is necessary to have available some universally acceptable language and communication system. In chemistry, this system consists of a language of mathematics, by which the quantitative aspects of the science are described, and a language of special terms and phrases, by which the qualitative aspects are described. The next chapter discusses units of measurement and the mathematical interpretation of data. The language of terms and phrases will be developed over the next several chapters as basic concepts of chemistry are introduced and discussed.

study questions

1 List five nonlaboratory processes or phenomena, other than those given on pages 5 and 7, that involve chemistry.

2 What are some of the subdivisions of chemistry? Why are these becoming less appropriate or adequate as subdivisions?

3 Why is chemistry referred to as an experimental science?

4 List the principal steps in the scientific method.

5 Outline the process by which you might determine the length of a day, using the scientific method.

6 What two properties are associated with the description of matter?

7 Describe how it could be shown that air is matter.

8 Compare the three states of matter in terms of the following characteristics.
   a. Distance between particles
   b. Definiteness of volume
   c. Definiteness of shape
   d. Relative degree of attractive forces between particles

9 Gases show the greatest degree of compressibility as compared to liquids and solids. Why?

10 Why does increasing the temperature of a gas in a closed container increase the observed pressure?

11 Suppose that you had a block of lead. What would you list as physical properties of that block?

12 How do the chemical properties of a substance differ from its physical properties?

13 Identify each of the following as either a chemical or a physical change.
   a. Making wine
   b. Generating electricity in a battery
   c. Vaporizing dry ice
   d. Making a milk shake
   e. Polishing a rock
   f. Igniting a flashbulb

14 Name some forms of energy.

special problems

1 Iron rusts by picking up oxygen from the air to form iron oxide; thus rust is actually iron oxide. Based on the Law of Conservation of Mass, how much iron oxide would be formed if 112 grams of iron react with 48 grams of oxygen? (Iron oxide is the only product.)

2 Propane is a fuel commonly used for cooking and heating in rural areas. Like natural gas, it burns to form carbon dioxide

and water. How much propane would be required to produce 132 grams of carbon dioxide and 72 grams of water if
160 grams of oxygen also were consumed from the air?

3 Identify the kinds and sequences of energy conversions
associated with each of the following processes.
a. Taking a picture in a camera
b. Detonating a stick of dynamite
c. Using an electrical burner to boil water
d. Growth of a green plant

4 Provide a brief explanation to justify each answer to Problem 3.

chapter **2**

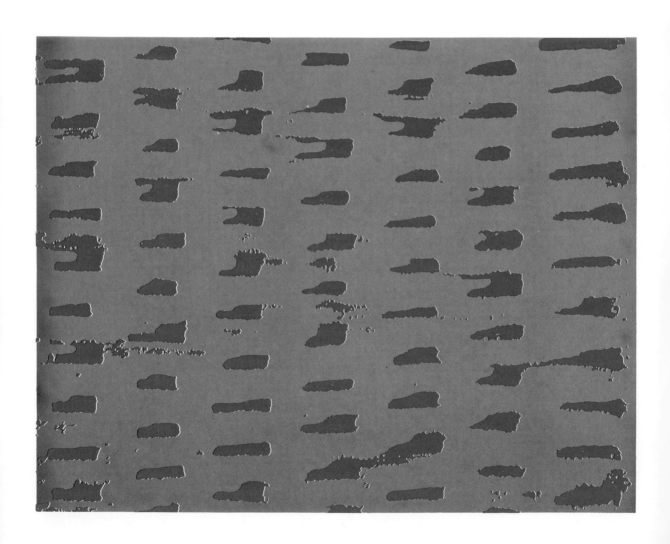

systems and units of measurement

Systems of measurement have existed probably for as long as civilized man. Descriptions of characteristics and observations are much more meaningful if expressed in precise units that are widely understood and accepted. To say that a box is heavy does not mean so much as saying that it weighs 26 pounds; to say that it is a warm day does not mean so much as saying that the temperature is 80°F. Doubt or ambiguity is unlikely to exist if precise units of measurement are used.

Establishment of any system of measurement is based on some arbitrarily selected set of basic units that is generally agreed on and accepted. For example, the common inch has been defined at one time as the average width of three thumbs—a large, a medium, and a small one—and at another time as the combined length of four barleycorns. The foot likewise has varied in length historically because of variations in the human feet selected as the basic reference. For certain measurements, such as those used for scientific purposes, however, a more precise, invariable set of standard units must be adopted.

Over the years, two independent systems of measurement have evolved. Presently in the United States, common nonscientific measurements, such as those done in the home or marketplace, are made by the so-called English system. This system uses the familiar units such as inch, foot, pound, and quart. Most countries, however, have adopted the metric system as their basic system of measurement, and this system is currently used in the United States for scientific work.

The United States is considering a recommendation to convert all measurements to the metric system, a conversion Great Britain is now making. Some of the major benefits and difficulties associated with such a conversion will be discussed at the end of this chapter.

English system of
  measurement
metric system of
  measurement
International System of Units
  (SI)
meter
gram
mass
weight
liter
heat
temperature
calorie
Calorie
joule
Fahrenheit
Celsius (centigrade)
(Kelvin) absolute
absolute zero
density
specific gravity

metric system

The metric system is based on a decimal plan, in which larger units differ by multiples of 10 and smaller units by multiples of ¹/₁₀. The system originated in France in 1798, as an attempt to better organize the nations of the world with one universal system of measurement. In 1866, it was approved by Congress for use in the United States, but it did not replace the English system already prevailing.

In 1960, an extension of the metric system, known as the International System of Units (commonly referred to by the initials SI, which stand for *Système Internationale*) was proposed by the International Bureau of Weights and Measures. Its use has been endorsed by major scientific organizations around the world. This modification of the metric system is a further attempt to develop a universal system to facilitate communication of numerical data. Its distinctive features include redefinition of the basic units of measurement in more precise terms and suggestions of new expressions of values to replace some presently used units.

Ancient length standard, the royal cubit of Egypt (about 1550 BC), made of black granite. The cubit (which means *forearm*) was subdivided into 2 spans, 6 palms, and 24 digits or finger breadths. (*Courtesy National Bureau of Standards*)

metric units and prefixes

The basic units of length, mass, and volume in the metric system are the meter, the gram, and the liter respectively. A meter is a little longer than a yard, a gram is about ¹/₂₈ of an ounce, and a liter is nearly equal to a quart.

When the metric system was proposed, the basic units were defined in the following way. A meter was said to be equal to one ten-millionth of the length of a meridian of the earth between the equator and the North Pole. The actual measurement was made on the meridian connecting Dunkirk and Barcelona. This standard was preserved by marking the appropriate length on a bar of platinum at the temperature of melting ice. This bar then was deposited in the Archives of the Republic of France. Exact replicas in a platinum-iridium alloy have been made available to other nations. The United States' replica is kept by the National Bureau of Standards.

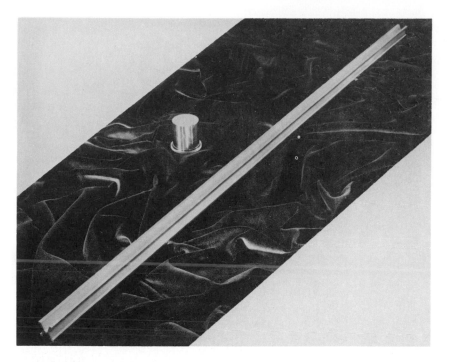

The United States standards for the kilogram and the meter. These exact duplicates of the original international standards kept in Paris are made of a stable alloy of 90% platinum and 10% iridium. The meter bar has been replaced as the standard of length by the wavelength of the orange-red light emitted by an isotope of krypton. The meter bar is still important, however, because of its suitability for certain types of measurements. (*Courtesy National Bureau of Standards*)

In 1960, when the International System was proposed, the meter was redefined more precisely as 1,650,763.73 wavelengths in a vacuum of the light responsible for the orange-red line in the spectrum of krypton-86. The new standard was chosen because the wavelength of light is believed to be independent of all known variables, such as temperature and humidity.

(*Courtesy National Bureau of Standards*)

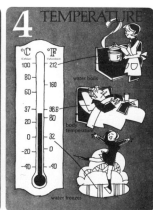

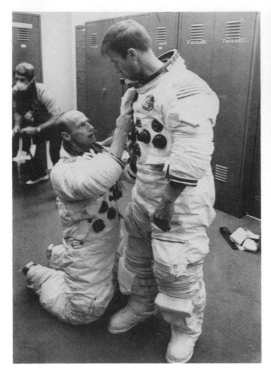

A man on earth (*left*) having a mass of 200 pounds also weighs 200 pounds. Standing on the moon (*right*), he still has a mass of 200 pounds, but a weight of only about 34 pounds. (Courtesy *left*, UPI Photo; *right*, NASA)

The metric unit of mass, the gram, originally was based on the meter; the gram was defined as the mass of a cube of water one-hundredth of a meter on an edge at the temperature of greatest density, 4°C. Because this mass was relatively small, preservation of this standard was made with a cylinder of platinum weighing as precisely as possible a thousand times this amount, or one kilogram. This standard also was deposited in the French Archives. The gram presently is defined as one-thousandth of the weight of the standard platinum-iridium kilogram.

It should be noted that the terms *mass* and *weight* actually refer to two different properties. Mass refers to the total quantity of matter possessed by an object; weight refers to the gravitational attraction between the earth and that object. The mass of a substance does not vary, but its weight may, depending on the substance's location relative to the earth or some similar body. For example, a man standing on the moon will have the same mass as he does on earth, but his weight will be only about ⅙ as much as his weight on earth. Because the gravitational attraction of the earth for an object on or near the surface does not vary appreciably from one location to another, mass and weight often are used interchangeably.

The liter, the most common standard unit of volume, was defined originally as the volume occupied by 1000 grams of water at 4°C. The new International System, however, defines it as one-thousandth of the volume of a cubic meter (0.001 m^3).

Designations of very large or very small numbers often are made in exponential form, using powers of 10. For example, the number 100 can be represented by 1×10^2, or simply 10^2. The designation 10^2 means 10×10. Likewise, 1000 can be represented as 1×10^3, and 4300 as 4.3×10^3. The designation 10^0 represents 1. Numbers less than 1 also can be designated in exponential form, using negative powers of 10. Thus, 1×10^{-1} equals $1 \times \frac{1}{10}$ or 0.1, and 1×10^{-2} equals $1 \times \frac{1}{10} \times \frac{1}{10}$, or $\frac{1}{100}$, or 0.01. Similarly, 7.5×10^{-4} represents 0.00075 ($7.5 \times \frac{1}{10} \times \frac{1}{10} \times \frac{1}{10} \times \frac{1}{10}$, or $7.5 \times \frac{1}{10,000}$).

Other units in the metric system are derived from the basic units by using appropriate prefixes that denote different magnitudes. Some of the more commonly used prefixes and their meanings are given in Table 2-1.

Table 2-1. Prefixes used in the modernized metric system

mega (M) = one million, or 10^6, times a unit
kilo (k)   – one thousand, or 10^3, times a unit
hecto (h) = one hundred, or 10^2, times a unit
deka (da) = ten, or 10^1, times a unit
deci (d)   = one-tenth, or 10^{-1}, of a unit
centi (c)   = one-hundredth, or 10^{-2}, of a unit
milli (m)   = one-thousandth, or 10^{-3}, of a unit
micro (μ) = one-millionth, or 10^{-6}, of a unit

The use and the relationships created by these prefixes are illustrated in Table 2-2.

Table 2-2. Examples of relationships between metric units

1,000,000 liters = 1 megaliter (Ml)
1,000 liters = 1 kiloliter (kl)
100 liters = 1 hectoliter (hl)
10 liters = 1 dekaliter (dal)
1 liter = 10 deciliters (dl)
1 liter = 100 centiliters (cl)
1 liter = 1,000 milliliters (ml)
1 liter = 1,000,000 microliters (μl)

conversions and calculations

Mathematical manipulations concerning the metric system involve conversions of two types—conversions *within* the metric system and conversions *between* the metric system and the English system.

Examples of conversions within the metric system are given below.

> 1 A metal can holds 3.79 liters of gasoline. To how many milliliters is this equal?
> a. There are 1000 milliliters (ml) in 1 liter.
> b. Therefore, in 3.79 liters, there are 3.79 times 1000 milliliters or
>
> $$3.79 \text{ l} \times \frac{1000 \text{ ml}}{\text{l}} = 3790 \text{ milliliters}.$$
>
> 2 A man is 175 centimeters tall. What is his height in meters?
> a. There are 100 centimeters (cm) in 1 meter.
> b. Therefore, 175 centimeters should equal $175/100$ meters, or
>
> $$\frac{175 \text{ cm}}{100 \text{ cm/m}} = 1.75 \text{ meters}.$$
>
> c. Since each centimeter is equivalent to 0.01 meter, the problem also could be solved by:
>
> $$175 \text{ cm} \times \frac{0.01 \text{ m}}{\text{cm}} = 1.75 \text{ meters}.$$

Although solutions of conversion problems such as these may appear complex, they actually involve nothing more than a shift in the location of the decimal point. Ease of manipulation is an advantage of using the metric system of measurements.

Often there may be several ways to solve a mathematical problem. The choice of method is up to the individual. Observe that the manner in which the mathematical expressions are set up in these illustrations leads to an appropriate cancellation of units to yield the correct final answer. This technique often can be used to check a method of solving a problem. Finally, and most importantly, one should try to understand the reasoning process involved in solving these problems rather than memorize solution patterns, since not all problems are solved exactly the same way.

Since there are currently two major systems of measurement in practice, it is frequently necessary to convert between the systems, and certain conversion factors are necessary. Some of the frequently used conversion factors are given in Table 2-3.

Table 2-3. Conversion factors between metric and English units

2.54 centimeters = 1 inch	1 kilogram = 2.2 pounds
1 meter = 39.37 inches	1 liter = 1.06 quarts
1 kilometer = 0.62 miles	3.79 liters = 1 gallon
28.35 grams = 1 ounce	28.3 liters = 1 cubic foot
453.6 grams = 1 pound	

Examples of conversions between the metric system and the English system are illustrated below.

The measurements in centimeters of a young model are 91.5–61–89, and the model weighs 50.7 kilograms. What are her measurements in inches and her weight in pounds?

1 a. There are 2.54 centimeters in 1 inch.
   b. Therefore, 91.5 centimeters equal $^{91.5}/_{2.54}$

$$\frac{91.5 \text{ cm}}{2.54 \text{ cm/in}} = 36 \text{ inches.}$$

   c. Similarly, 61 centimeters and 89 centimeters equal 24 inches and 35 inches respectively.
   d. The model's measurements in inches are therefore 36–24–35.
2 a. There are 2.2 pounds in 1 kilogram.
   b. Therefore, 50.7 kilograms are equivalent to 50.7 times 2.2 pounds, or

$$50.7 \text{ kg} \times \frac{2.2 \text{ lbs}}{\text{kg}} = 111.5 \text{ pounds.}$$

Note that, in these examples, at least one appropriate conversion factor must be known and used. It is not necessary to learn many intersystem conversion factors so long as one is familiar with conversions within the metric and within the English systems themselves. It should be clear that such conversions are a nuisance and that they would be unnecessary if a system such as the metric were to be adopted universally as the principal system of measurements.

heat and temperature

As defined earlier, heat is due to the motion of particles in a substance. Temperature is a measurement of the intensity of this heat. For example, the heat evolved by a burning tree is much less than that evolved by a burning forest. Nevertheless, the temperature of the flames produced by the single burning tree could be the same as that of the flames produced by the burning forest.

The actual temperature of the flames in a single burning tree is approximately the same as the temperature of the flames in a burning forest, but the heat production of a single tree is much less than that of a forest. (*Courtesy U.S. Forest Service*)

Heat is expressed in terms of calories. A calorie is defined as the amount of heat energy involved in changing the temperature of one gram of water one degree Celsius (centigrade). This may be heat energy gained or lost by a substance. In practical applications, for example in determining the calorie content of certain foods, a unit larger than the calorie is desirable. For this purpose, the large Calorie (spelled with a capital C) is used. The large Calorie equals 1000 calories, or one kilocalorie.

Although the calorie is a widely used unit of heat, it has not been adopted as part of the International System of Units. The standard unit for energy, including heat, that has been approved in SI is the joule. One joule is approximately 0.24 calories.

The instrument used to measure the temperature of a system is the thermometer. In the United States, everyday measurements of temperature are made in degrees Fahrenheit (F). This scale was devised by Gabriel Fahrenheit, a German physicist, in the early 1700s. The two reference systems that he used were a salt/ice mixture and boiling mercury. He assumed these represented the greatest

The caloric content of various foods represents the heat-producing or energy-producing value of those foods. When food is burned metabolically in the body, it produces energy that the body uses for its functions. This energy is measured in calories. Certain types of foods, such as fats and starches, have higher caloric contents than other types of foods, such as proteins.

degrees of cold and heat and assigned readings of 0° to the salt/ice mixture and 600° to the boiling mercury. The distance between these two points on Fahrenheit's scale then was divided into 600 equal parts, or degrees. On this scale, water at sea level froze at 32° and boiled at 212°.

The temperature scale used most commonly in scientific work is the centigrade or Celsius scale (C). This scale was developed in 1742 by Anders Celsius, a Swedish astronomer. On the Celsius scale, water's freezing point is 0° and its boiling point is 100°.

For certain scientific measurements, particularly those involving gases, a third temperature scale is used. This scale is called the absolute or Kelvin scale (K). It was proposed by a British physicist, William Kelvin, in 1848. A Kelvin degree represents the same magnitude of heat as a Celsius degree, but the numbers on the temperature scales are different. Kelvin assigned the reading of 273° to the freezing point of water and the reading of 373° to its boiling point. The zero point on the Kelvin scale, now established accurately at −273.15°C, is termed absolute zero. Absolute zero is the temperature at which theoretically all thermal motion ceases in matter. Scientists have not yet produced absolute zero experimentally, although they have come within a fraction of reaching it. A comparison of the three temperature scales is provided in Figure 2-1.

Figure 2-1. Comparison of three temperature scales

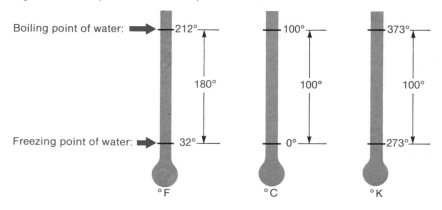

Mathematical formulas exist for converting a reading on one temperature scale to a reading on another scale.

1 To convert a Fahrenheit reading to a Celsius reading:

$$°C = \frac{°F - 32°}{1.8}$$

2 To convert a Celsius reading to a Fahrenheit reading:

$$°F = 1.8 \; °C + 32°$$

3 To correlate a Celsius reading to a Kelvin (absolute) reading:

$$°K = °C + 273°$$

density and specific gravity

It is generally accepted that aluminum is a lighter metal than lead. It is possible, however, for a piece of aluminum to weigh more than a piece of lead. What obviously is implied in the first statement is that the comparison is being made between samples of equal volume.

An expression of mass per unit volume is the density of a substance. Usually density is represented in grams per milliliter, although any unit of mass and any unit of volume may be used, so long as these are indicated clearly. For example, the density of water can be expressed as 1 gram/milliliter or 8.35 pounds/gallon. For gases, which are not very dense, the units more often used are grams per liter rather than grams per milliliter. The densities of some common materials are given in Table 2-4.

Density may be expressed mathematically as density = mass/volume, or $d = m/v$. If any two of these quantities are known, the third may be determined by proper mathematical manipulations.

Table 2-4. Densities of some common substances
(in g/ml at 20°C unless otherwise stated)

Gold	19.30	Glycerin	1.26
Lead	11.34	Olive oil	0.92
Aluminum	2.70	Ethyl alcohol	0.79
Salt	2.16	Carbon dioxide	1.97 g/l at 0°C
Sugar	1.59	Oxygen	1.43 g/l at 0°C
Chloroform	1.49	Air	1.29 g/l at 0°C

Densities of pure substances are distinguishing characteristics and can serve as means of identifying these substances. Furthermore, densities can indicate the purity or quality of substances.

Since density can be expressed in various units, the numerical value may differ, depending on the units used. To achieve a standard value that may be compared universally, specific gravity is used. The specific gravity of a substance is its density compared with that of water. It is calculated according to the formula below.

$$\text{specific gravity} = \frac{\text{density of substance}}{\text{density of water}}$$

In this comparison, it is essential that the density of water be expressed in the same units as the density of the object. Thus, if the density of the object is expressed in lbs/gal, then the density of water

The boy (*left*) lifts the "heavy" rock with ease—because it is low-density material used for movie stunts. The strongman (*right*) strains to lift a high-density dumbbell. (*Courtesy left, Universal Studios Tour; right, photograph by A. Marshall Licht.*)

also must be expressed in lbs/gal. Under these conditions, the specific gravity has no units and is said to be "dimensionless." When density is expressed in g/ml, the specific gravity is numerically equal to the density, since the density of water is assumed to be 1 g/ml. Specific gravity is the common unit found on devices that measure the charge of a battery or the protective value of antifreeze.

Although the mass of a substance does not change with temperature, its volume does, since heat affects the motion of particles and alters the distance between them. Consequently, the density or specific gravity of a substance varies with temperature. For example, water has a density of 1.00 g/ml only at 4°C. At higher and lower temperatures, it has a slightly lower density. Gases are especially prone to vary in density with temperature. For the most precise work, both units and temperature must be specified whenever density measurements are involved.

universal system of measurement

Since its origin in 1798, the metric system has been adopted by nearly every country. To accommodate the European Common Market, in 1965, Great Britain began a ten-year program to convert to the metric system. Both Australia and Canada have declared that they too will begin conversion programs. This will leave the United States as the only major industrial country that has not decided to switch to the metric system.

In 1968, the United States Congress ordered a study on the feasibility and desirability of conversion to the metric system. In 1971, the National Bureau of Standards recommended that the United States convert its system of measurements to the metric system. In making this recommendation, the Bureau recognized that there would be problems associated with the changeover, but felt that the many advantages of the new system would outweigh the disadvantages.

The benefits in the United States from the changeover can be enormous. Because the metric system is based on the decimal system, mathematical manipulations will be easier and errors in calculations will be less frequent. Cumbersome conversions between the English and metric systems of measurement will no longer be needed. This will eliminate, for instance, a problem that engineers encounter in commercializing or industrializing a discovery that has been developed in a laboratory oriented to metric measurements. Replacement parts for foreign-made goods can be easier to obtain, because

As one way of integrating the metric system into American life, traffic signs showing both English and metric units now occur on highways. (*Courtesy Huntsville Department of Transportation*)

the parts could be manufactured in the United States if factory equipment were standardized to the metric system. The most significant benefit of metrication will probably be in world trade, however. After World War II, the United States was one of the few major countries that still had an intact industrial complex. For the next twenty-five years, the United States dominated world trade. More recently, the industrial capacity of other major nations has recovered completely. With increased cooperation among the European Common Market countries, the development of Japan as an economic power, and worldwide standardization via the metric system, the United States can expect to find its "oddball-sized" products increasingly bypassed in world trade if it does not convert to the metric system.

Several government and commercial concerns already have made or are making the change. For many years the medical profession and the pharmaceutical industry have used only the metric system, and since 1970, NASA has been exclusively metric in its measurement system. The textile industry, the Ford Motor Company, the General Motors Corporation, the IBM Corporation, and the Caterpillar Tractor Company gradually are converting to metric measurements for their new products. Traffic signs showing speed limits in both miles per hour and kilometers per hour are beginning to appear on the nation's highways.

For metrication to be most successful, however, a logical, coordinated plan is necessary. Such a plan will necessarily involve a gradual, mandatory changeover accompanied by education of the public. Education probably will be accomplished best by instructing the young through the schools and by informing adults by quoting weights and measures in both systems for a time. Both education processes are already in use. In some elementary schools, children are being introduced to the metric system. The labels of some commercial products, such as cereals, canned fruits, and vegetables, now express weights and volumes in both English and metric units. Furthermore, international track and field, swimming, and diving events are measured in meters.

The changeover will not be easy, however, and the cost in time, effort, and money will be great. Every item manufactured in the United States must be made to a new standard of measurement. Even common items such as nuts and bolts will be different. A standard 1/4-inch bolt has 20 threads per inch. The nearest metric equivalent is a 6-millimeter bolt with 10 threads per centimeter. These bolts and their corresponding nuts are not interchangeable. Furthermore, the wrenches required for these two bolt sizes are different, and existing wrenches will not fit the new sizes. Auto mechanics discovered this in the early 1950s, when foreign cars first became popular in the United States and mechanics were unequipped to work on them.

The dimensions of every man-made object can be expected to change to accommodate the new system of measurement. The dimensions of a typical city block and street will change because of the measurements of length in meters instead of feet. Door and window sizes will be modified, as well as plumbing fittings and electrical appliances. Housewives will buy a liter of milk and half a kilogram of fruit rather than the now-familiar quart or pound.

Most significantly affected will be industries dealing with the manufacture of machinery, instruments, and maintenance equipment. Commonly, all specifications of industrial equipment and products have been expressed in English units. The change to the metric system will necessitate replacing all manufacturing and maintenance equipment in individual plants. The replacement must be gradual, not abrupt, because products of these industries have wide uses by many individuals and smaller firms. For a time, dual standards and stocking of supplies will be necessary to permit a smooth transition between systems.

In spite of the problems associated with the changeover, the trend toward worldwide metrication has already gone so far that the issue is no longer *whether* the United States should convert to the metric system. The only sensible question that now may be asked is

"When and how can the change be made?" Gallup polls taken in the last decade show an increasing awareness of the metric system by the general public. Those favoring adopting this system as the standard of measurement also have increased in percentage in the last decade. In 1965, 37% of the public favored adoption; in 1971, 52%; in 1973, 55%. The United States Congress has been slow to act, however. As recently as May, 1974, the House of Representatives rejected a bill to bring about the changeover on a voluntary basis over a ten-year period, because the issue of who will pay for the change could not be resolved. When the United States finally does make the inevitable change, the metric system will indeed have become a universal system of measurement.

study questions

1 Comment on the origin of systems of measurement.
2 Which system of units is more accurate, the English system or the metric system?
3 Which system is easier to use, the English system or the metric system?
4 What are the *standard* metric units of measurement of length, volume, and mass?
5 What information is provided by a prefix such as *milli* or *deci* in the name of a metric unit of measurement?
6 Is a block of aluminum always lighter than a block of iron?
7 The mass of an object does not vary, but its weight may. Why?
8 What is the difference between the density of an object and its specific gravity? Are they always numerically equal?
9 How would you determine the density of a coin?
10 The temperature of a match flame is not much different from that of a log burning in a fireplace, but more heat is evolved from the fireplace. Comment on this apparent contradiction.
11 What are the differences among the Fahrenheit, Celsius, and Kelvin temperature scales?
12 Why would a universal system of measurement be desirable?
13 As the United States undertakes conversion of its system of measurement to the metric system, what difficulties do you expect will be encountered, and what solutions can you offer to overcome these difficulties?

special problems

1 Make the following conversions.
   a. 25 pints to gallons
   b. 42 inches to yards
   c. 198 ounces to pounds
   d. 467 yards to miles
   e. 0.0379 meter to centimeters
   f. 427 meters to kilometers
   g. 8.45 liters to milliliters
   h. 3.54 kilograms to grams
   i. 5.5 pounds to grams
   j. 67.3 centimeters to yards
   k. 3.25 pints to liters
   l. 4 ounces to grams

2 A sample of alcohol weighed 23.7 grams. It was poured into a graduated cylinder and found to have a volume of 30 milliliters.
   a. What is the density of the alcohol?
   b. What is the specific gravity of the alcohol?

3 A block of lead weighed 47 grams. When placed in a graduated cylinder containing 25 milliliters of water, the final volume was 29.14 milliliters.
   a. Determine the density of the block of lead.
   b. Determine the specific gravity of the block of lead.

4 The low temperature in some parts of the country reaches 10°F during the winter.
   a. What is the corresponding Celsius temperature (°C)?
   b. What is the corresponding absolute temperature (°K)?

5 Dry ice (solid carbon dioxide) vaporizes at −40°C.
   a. What is the corresponding Fahrenheit temperature?
   b. What is the corresponding absolute temperature?

chapter **3**

elements, compounds, and mixtures

It was stated in Chapter 1 that elements and compounds are types of pure matter, whereas mixtures are impure matter. The terms *pure* and *impure* refer to composition, or more specifically to constancy of composition. Pure matter, such as an element or a compound, shows no variation in its chemical composition, regardless of the source of a sample. This composition is also definite for and unique to a given substance and distinguishes a substance from all others. If matter has a variable and indefinite composition, it is termed a *mixture* and is considered impure. This chapter will explore more fully these and other distinguishing characteristics of elements, compounds, and mixtures.

atom
chemical symbol
Law of Constant Composition
Law of Definite Proportions
chemical formula
subscript
molecule
chemical equation
coefficient
balanced chemical equation

elements

Presently, there are 105 known elements. These substances are the simplest forms of matter, each possessing distinguishing chemical characteristics unique to itself. They cannot be broken down into simpler matter, or other elements, by ordinary chemical means. Because of these known scientific facts, elements have been called the "building blocks" of matter.

Not all 105 known elements occur naturally. There is no agreement among authorities as to the exact number, but it is generally accepted that about 90 elements occur in nature, either as the pure element or more commonly as a component in compounds. The remaining 15 or so are artificial, or synthetic, elements, which are products of man-induced processes. Continuing research can be expected to produce additional man-made elements. These synthetic elements are all characteristically unstable. Although they have not been found in nature, it is conceivable that in the early history of the earth, these and possibly other unknown elements may have been present. Their unstable nature could have resulted in their disappearance from the natural components of the earth, however.

The naturally occurring elements were not all discovered at the same time. Some, such as gold, silver, iron, lead, and copper, have been known since the days of earliest civilizations. Others, such as helium, radium, aluminum, and bromine, were discovered in the nineteenth century. The most abundant elements found in the earth's crust, in order of decreasing percentage, are oxygen, silicon, aluminum, and iron. Others present in amounts of 1% or more are calcium, sodium, potassium, and magnesium. Together, these represent about 98.5% of the earth's crust.

Representative collection of elements including metals, nonmetals, solids, liquids, and gases (in cylinders). (*Alfa Products catalog cover reproduction with permission of Ventron Corporation, Alfa Products*)

The first artificially produced element was technetium. It was synthesized in 1937, by scientists at the University of California at Berkeley who used a cyclotron to change another element, molybdenum, into this new one. Since then, other synthetic elements have been produced by similar methods.

The fundamental particle of matter found in elements is the **atom.** The atom is the smallest particle of an element that still bears all the chemical characteristics associated with that element. Although atoms do consist of smaller, subatomic particles, none of these particles are characteristic in themselves of any one element, and atoms cannot be broken down into simpler atoms by ordinary chemical means. All atoms of a given element have similarities in their compositions that identify them as atoms of the same element. Atoms of two or more different elements do not share this similarity in composition. Further discussion of these characteristic compositions will be given in Chapter 4.

The names of the elements have been derived from a variety of sources. Some of these are outlined below.

Names derived from outstanding properties or characteristics:

Iodine — from the Greek *iodes*, meaning violet

Platinum — from the Spanish *plata*, meaning silvery

Hydrogen — from the Greek *hydro genes*, meaning water former

Names representing sources of the elements:

Carbon — from the Latin *carbo*, meaning coal

Helium — from the Greek *Helios*, meaning the sun, where helium was first detected

Calcium — from the Latin *calx*, meaning lime

Names representing certain countries or regions:

Francium — for France, the native country of its discoverer, Marguerite Perey

Polonium — for Poland, the native country of its discoverer, Madame Curie

Americium — for America, where it was first produced

Names honoring certain individuals:

Einsteinium — for Albert Einstein, world-renowned mathematician and theoretician

Nobelium — for Alfred Nobel, who made contributions to the advancement of science

Niobium — for Niobe, the daughter of Tantalus in Greek mythology, since this element is always found associated with the element tantalum in nature

Names of miscellaneous origins:

Technetium — from the Greek *technikos*, meaning an art, since this was the first synthetically produced element

Neptunium — named after the planet Neptune, the first planet beyond Uranus, since neptunium is the first element beyond uranium

Tantalum — for Tantalus of Greek mythology, because the metal was hard to isolate and therefore tantalizing

The Alchemist, by David Teniers the Younger (1610–1690), depicts the paraphernalia probably used by alchemists in their attempts to make gold and silver from the common metals. (*Courtesy Fisher Collection, Fisher Scientific Company*)

Figure 3-1.   Some alchemical symbols

chemical symbols

In scientific writings of all kinds, from laboratory reports to textbooks and published research articles, it is desirable to simplify the communication process as much as possible. Systems of shorthand or abbreviations have been developed to help simplify the representation of scientific findings and ideas. The shorthand notation accepted for an element is called its chemical symbol.

The concept of using a chemical symbol to represent a chemical substance goes back to medieval times, when alchemists were trying to find ways of converting the common base metals to the more precious ones. Alchemists developed a system of figures to represent elements with which they were familiar. Rather than being a system of shorthand for ease of recording data, it is more likely that these symbols were invented to keep the work and findings of the alchemists secret, since the rewards of discovering the magic process were expected to be high. Some alchemical symbols are illustrated in Figure 3-1.

The modern system of chemical symbols is based on letters and is credited to a Swedish chemist, J. J. Berzelius. Developed by Berzelius in the early nineteenth century, the system derives chemical symbols for the elements in several ways. Single-letter chemical symbols such as C for carbon or H for hydrogen are derived from the first letter of the anglicized names of the elements. Because a number of elements have names that begin with the same letter, double-letter chemical symbols are also necessary, such as Co for cobalt, Cr for chromium, and He for helium. Note that these may be derived from either the first and second or the first and third letters of the anglicized names. Occasionally, a chemical symbol has been derived from the original Latin name of an element rather than from its present, anglicized one. The following chemical symbols are examples of these derivations: Ag for silver, from *argentum*, Na for sodium, from *natrium*, and Fe for iron, from *ferrum*.

For correct and unambiguous representations of elements by their accepted chemical symbols, certain rules must *always* be followed. Single-letter chemicals symbols are *always* written in capitalized form. Double-letter chemical symbols are written with the first letter *always* capitalized and the second letter *always* in lower case. Nonadherence to these rules creates confusion by suggesting representations of compounds rather than of elements.

In addition to generally representing the element itself, a chemical symbol also may represent one atom of a particular element and, in certain situations, a given quantity or weight of that element. The chemical symbols of all elements are given in Table 3-1.

Table 3-1. Chemical symbols

Element	Symbol	Atomic Number	Atomic Weight	Element	Symbol	Atomic Number	Atomic Weight
Actinium	Ac	89	(227)	Mendelevium	Md	101	(256)
Aluminum	Al	13	26.98	Mercury	Hg	80	200.59
Americium	Am	95	(243)	Molybdenum	Mo	42	95.94
Antimony	Sb	51	121.75	Neodymium	Nd	60	144.24
Argon	Ar	18	39.95	Neon	Ne	10	20.18
Arsenic	As	33	74.92	Neptunium	Np	93	237.05
Astatine	At	85	(210)	Nickel	Ni	28	58.71
Barium	Ba	56	137.34	Niobium	Nb	41	92.91
Berkelium	Bk	97	(247)	Nitrogen	N	7	14.01
Beryllium	Be	4	9.01	Nobelium	No	102	(254)
Bismuth	Bi	83	208.98	Osmium	Os	76	190.2
Boron	B	5	10.81	Oxygen	O	8	16.00
Bromine	Br	35	79.90	Palladium	Pd	46	106.4
Cadmium	Cd	48	112.40	Phosphorus	P	15	30.97
Calcium	Ca	20	40.08	Platinum	Pt	78	195.09
Californium	Cf	98	(249)	Plutonium	Pu	94	(242)
Carbon	C	6	12.01	Polonium	Po	84	(210)
Cerium	Ce	58	140.12	Potassium	K	19	39.10
Cesium	Cs	55	132.91	Praseodymium	Pr	59	140.91
Chlorine	Cl	17	35.45	Promethium	Pm	61	(147)
Chromium	Cr	24	52.00	Protactinium	Pa	91	231.04
Cobalt	Co	27	58.93	Radium	Ra	88	226.03
Copper	Cu	29	63.55	Radon	Rn	86	(222)
Curium	Cm	96	(247)	Rhenium	Re	75	186.2
Dysprosium	Dy	66	162.50	Rhodium	Rh	45	102.91
Einsteinium	Es	99	(254)	Rubidium	Rb	37	85.47
Erbium	Er	68	167.26	Ruthenium	Ru	44	101.07
Europium	Eu	63	151.96	Samarium	Sm	62	150.4
Fermium	Fm	100	(253)	Scandium	Sc	21	44.96
Fluorine	F	9	19.00	Selenium	Se	34	78.96
Francium	Fr	87	(223)	Silicon	Si	14	28.09
Gadolinium	Gd	64	157.25	Silver	Ag	47	107.87
Gallium	Ga	31	69.72	Sodium	Na	11	22.99
Germanium	Ge	32	72.59	Strontium	Sr	38	87.62
Gold	Au	79	196.97	Sulfur	S	16	32.06
Hafnium	Hf	72	178.49	Tantalum	Ta	73	180.95
*Hahnium	Ha	105	(260)	Technetium	Tc	43	98.91
Helium	He	2	4.00	Tellurium	Te	52	127.60
Holmium	Ho	67	164.93	Terbium	Tb	65	158.93
Hydrogen	H	1	1.01	Thallium	Tl	81	204.37
Indium	In	49	114.82	Thorium	Th	90	232.04
Iodine	I	53	126.90	Thulium	Tm	69	168.93
Iridium	Ir	77	192.22	Tin	Sn	50	118.69
Iron	Fe	26	55.85	Titanium	Ti	22	47.90
Krypton	Kr	36	83.80	Tungsten	W	74	183.85
*Kurchatovium	Ku	104	(257)	Uranium	U	92	238.03
Lanthanum	La	57	138.91	Vanadium	V	23	50.94
Lawrencium	Lr	103	(257)	Xenon	Xe	54	131.30
Lead	Pb	82	207.2	Ytterbium	Yb	70	173.04
Lithium	Li	3	6.94	Yttrium	Y	39	88.91
Lutetium	Lu	71	174.97	Zinc	Zn	30	65.37
Magnesium	Mg	12	24.31	Zirconium	Zr	40	91.22
Manganese	Mn	25	54.94	*Unofficial name and symbol			

compounds

When zinc metal is placed in hydrochloric acid, the resulting chemical reaction yields a new compound. The reaction can be represented by the following:

$Zn + 2 HCl \rightarrow ZnCl_2 + H_2$ (gas)

(*Courtesy International Nickel Company, Inc.*)

When two or more elements undergo chemical interaction with one another, compounds result. In such chemical changes, the elements involved lose the physical and chemical characteristics by which they normally are identified. The resulting compound possesses different composition and properties.

The elements involved always combine in a definite ratio of atoms when forming a particular compound, so compounds are said to have definite and constant compositions. For example, carbon dioxide always is formed by the chemical combination of carbon and oxygen in a 1 to 2 atomic ratio, but ammonia always is represented by a 1 to 3 atomic ratio of nitrogen and hydrogen. Changing either the ratio or a component element changes the composition of a substance, and the compound is no longer the same. For example, a 1 to 1 atomic ratio of carbon and oxygen represents carbon monoxide, a

Below, left: Napthalene sublimes when heated above room temperature. *Right:* When heated to high temperatures in air, naphthalene burns to produce black solid particles and some gas. As a result of the chemical reaction, new substances have been formed. (*Used with permission of Earlham College Press, copyright owner, and McGraw-Hill Book Company, publisher.*)

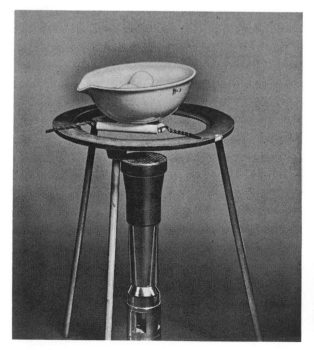

substance quite different from carbon dioxide in both composition and properties. These concepts are formalized in the Law of Constant Composition and in the related Law of Definite Proportions, which states that when elements combine to form a compound, they do so in a fixed and definite proportion by weight.

Component elements in compounds can be separated only by chemical means, not physical. Therefore, unlike elements, compounds can be broken down by ordinary chemical processes into simpler substances. These substances may be either simpler compounds or the component elements themselves.

Over two million compounds are known. Usually these are considered in two groups, inorganic and organic compounds. The larger of the two groups is the organic, which includes the majority of the carbon-containing compounds. The division is not a rigid or critical one, since the principles that pertain to the chemistry of one group are not distinctly different from those that pertain to the other group.

chemical formulas

A system of chemical notation has been established for compounds, similar in purpose to that serving for elements. The accepted short-hand representation of a compound is called its chemical formula.

For a chemical formula to be correct, it must contain two pieces of information: (1) it must indicate the elements in the makeup of the compound, and (2) it must indicate the combining ratio of atoms of these elements in the particular compound. These two qualifications of a correct chemical formula are necessary because of the Law of Constant Composition. The first piece of information is provided by including in the chemical formula correct chemical symbols for all elements in the compound. The second piece of information is provided by subscripts, numbers written to the right of and slightly below the chemical symbols of the elements. The subscript 1 is not indicated, but is understood where needed. The following examples illustrate these concepts.

Sulfuric acid, a common laboratory acid, is a combination of hydrogen, sulfur, and oxygen, giving a base formula of HSO. These elements combine in the atomic ratio of 2 to 1 to 4. Therefore, the correct chemical formula for sulfuric acid is H_2SO_4.

The correct chemical formula for the compound sodium hydrogencarbonate (bicarbonate of soda) is $NaHCO_3$. The elements composing this substance are sodium, hydrogen, carbon, and oxygen in the combining ratio of 1 to 1 to 1 to 3.

Occasionally, the chemical formula for a compound will contain parentheses, such as $Ca(NO_3)_2$. This provides the information that a particular combination of atoms exists as a distinct subunit within the chemical makeup of the compound. In this example, the combination NO_3 represents the nitrate unit, and the formula can be interpreted to mean that each calcium unit is combined chemically with two nitrate units. The chemical formula also can be interpreted correctly in terms of the two basic qualifications listed earlier: the elements involved are calcium, nitrogen, and oxygen in the atomic ratio of 1 to 2 to 6. Note that, in this analysis, the subscript 2 pertains to everything within the parentheses and therefore doubles the number of nitrogen and oxygen atoms shown within the parentheses. The chemical formulas for a few common substances are given in Table 3-2.

Table 3-2. Names and formulas of some common chemical compounds

Common Name	Chemical Name	Chemical Formula
Limestone	Calcium carbonate	$CaCO_3$
Baking soda	Sodium hydrogencarbonate	$NaHCO_3$
Washing soda	Sodium carbonate	Na_2CO_3
Battery acid	Sulfuric acid	H_2SO_4
Table salt	Sodium chloride	$NaCl$
Sugar	Sucrose	$C_{12}H_{22}O_{11}$
Aspirin	Acetylsalicylic acid	$C_9H_8O_4$

The fundamental particle of matter in most compounds is the **molecule,** which results from the chemical interaction of two or more atoms. The molecule is the smallest particle of matter that contains all of the properties of a compound and that contains the same ratio of atoms of the component elements as found in the compound itself.

Relatively fewer compounds are composed of ions rather than of molecules. Ions differ from atoms and molecules by being electrically charged particles of matter. The charges may be either positive or negative and generally vary in magnitude from one to three. The nitrate unit mentioned earlier is an ion. The calcium unit with which the nitrate is associated in the compound $Ca(NO_3)_2$ is also an ion. When ions constitute the fundamental particles of matter in a compound, they occur in such numbers that the total positive charge equals the total negative charge, and an overall electrically neutral substance results.

$$Ca(NO_3)_2 = Ca^{+2} + 2\ NO_3^{-1}\ \text{(Net charge is zero.)}$$

mixtures

If two or more substances (elements or compounds) are brought together, without a chemical interaction occurring between them, the resulting system is called a *mixture*. Mixtures differ from compounds in that the components of a mixture do not lose their original identities and characteristics and generally can be separated by physical processes and without chemical reactions.

Consider, for example, a mixture of salt and pepper. Close examination will show distinct particles of salt and of pepper with no obvious formation of a new substance. The salt particles are identical in properties and composition to the particles in pure salt, and the pepper particles in the mixture are identical in every respect to particles in a can of ground pepper. These two components can be separated either manually, by picking out all salt grains, or by dissolving the salt particles in water, removing the insoluble pepper particles, and evaporating the water to recover the salt. Other examples of this type of mixture are muddy water and smoke.

Mixtures need not have visually distinguishable components so long as it can be shown that the components have retained their individual characteristics, on being physically brought together. Mixtures of this type, such as salt water, ink, or gasoline, are solutions and are physically homogeneous because all parts appear to be identical. Chemically, however, such mixtures are heterogeneous, because not all units of matter within the mixture are chemically identical. The salt and pepper mixture would be characterized as physically as well as chemically heterogeneous. Elements and compounds are chemically homogeneous, but may exist in physically homogeneous or heterogeneous situations. For example, pure water is both chemically and physically homogeneous. A glass of water containing chips of ice, however, would be chemically homogeneous, but physically heterogeneous, because of the presence of two different phases or states of matter. When one speaks of homogeneity or heterogeneity of matter, it is helpful to specify whether the terms refer to chemical composition or physical existence.

Since no chemical interaction or binding exists between components in a mixture, it is impossible to define precisely the composition of any mixture in terms of combined ratios of atoms. Mixtures will vary in composition from area to area within a given sample, and from sample to sample as well. Air is a mixture of gases, and although samples of air taken from different locations usually will agree in principal components that are present (such as oxygen, water vapor, carbon dioxide, and nitrogen), the samples are unlikely to agree in exact percentage compositions. For instance, a sample of

Magnified crystals of sucrose (*top*), sodium bromide (*center*), and a mixture of sucrose and sodium bromide (*bottom*). The mixture is physically and chemically heterogeneous, but individually the sucrose and the sodium bromide are homogeneous. (*Used with permission of Earlham College Press, copyright owner, and McGraw-Hill Book Company, publisher.*)

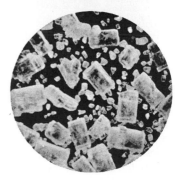

Gold bars in a bank vault. The pure element is both physically and chemically homogeneous. (*UPI Photo*)

Magnified crystals of salt (sodium chloride), a physically and chemically homogeneous substance. (*Courtesy Morton-Norwich Products, Inc.*)

air taken from a desert region is likely to have a lower water vapor concentration than a sample taken from a tropical jungle. The Law of Constant Composition does not apply to mixtures; consequently no chemical formulas can be written for them because of their variable and indefinite compositions.

chemical equations

Two types of chemical notation or shorthand already have been discussed—the chemical symbol and the chemical formula. One more type of notation is available to scientists. This is the chemical equation, which is the representation of a chemical reaction. To be correct, a chemical equation not only must contain representations of all starting materials and final products, but also must indicate the relative amounts of the substances that participate in the reaction. In this way, the chemical equation correctly illustrates the conservation of mass, which the reaction follows.

Participating substances are represented in a chemical equation by their correct symbols and formulas. The relative amounts of material involved in the reaction are indicated by **coefficients**, numbers that are placed before symbols and formulas in the chemical equation. When this is done, the equation is said to be a **balanced chemical equation**. These numbers, or coefficients, represent multiples of units of elements and compounds that are part of a particular chemical equation. When these coefficients are properly provided, the chemical equation is balanced. The distinction between a coefficient and a subscript is shown below.

$$\text{coefficient} \rightarrow 2\ HNO_3 \leftarrow \text{subscript}$$

(applies to the entire formula)   (applies only to the oxygen)

The following examples illustrate some techniques involved in writing and balancing chemical equations.

1 Combustion in air of methane, a principal constituent in natural gas, produces carbon dioxide and water.

2 Correct representations of the substances in this reaction are: CH_4, methane; O_2, oxygen gas from the air, CO_2, carbon dioxide; and H_2O, water.

3 The basic or unbalanced equation for this combustion process would be $CH_4 + O_2 \rightarrow CO_2 + H_2O$. The plus sign means "combines or reacts with," and the arrow means "to yield or to produce." Therefore, the equation would be read as "methane combines or reacts with oxygen gas to produce carbon dioxide and water."

4 The equation must now be balanced to adhere to the Law of Conservation of Mass. Balancing a chemical equation involves equating the number of atoms of each element on both sides of the arrow. Presently, the numbers of carbon atoms on both sides of the chemical equation are equal (one atom on each side), and carbon is considered balanced.

5 At present the numbers of hydrogen atoms are not equal on both sides, since there are four on the left and only two on the right. To balance the hydrogen atoms, a coefficient of 2 must be placed before the formula for water on the right.

$$CH_4 + O_2 \rightarrow CO_2 + 2\ H_2O$$

Since each water molecule contains two hydrogen atoms, the coefficient of 2 that doubles the number of water molecules also provides a total of four hydrogen atoms on the right side of the chemical equation. The hydrogen atoms are now balanced. It is not acceptable to balance the hydrogen atoms by changing the formula H_2O to H_4O. Although this provides the four hydrogens theoretically needed on the right, it also changes the composition of one of the given substances, since H_2O and H_4O do not represent the same compound. Consequently, balancing equations cannot be done by changing subscripts.

6 To balance the oxygen atoms involved on both sides of the equation, a coefficient of 2 must be placed before the formula for oxygen gas.

$$CH_4 + 2\,O_2 \rightarrow CO_2 + 2\,H_2O$$

This provides four oxygen atoms on each side of the chemical equation, which is now completely balanced.

Note that in the last balancing step above, both CO_2 and H_2O were considered when determining the total number of oxygen atoms on the right side of the chemical equation. This procedure is always necessary whenever an element appears in more than one substance on one or both sides of a chemical equation, so that a correct count of atoms of that element is obtained. Generally speaking, the balancing process often can be simplified if those elements, such as oxygen and hydrogen, that are widely distributed among members of a chemical equation are balanced last. Note the following sequence of steps, in which balancing of oxygen and hydrogen is not attempted initially.

$$Ca(OH)_2 + HNO_3 \rightarrow Ca(NO_3)_2 + H_2O$$
(Ca already balanced.)

$$Ca(OH)_2 + 2\,HNO_3 \rightarrow Ca(NO_3)_2 + H_2O$$
(To balance N, or NO_3.)

$$Ca(OH)_2 + 2\,HNO_3 \rightarrow Ca(NO_3)_2 + 2\,H_2O$$
(To balance H and O. Entire equation now balanced.)

Occasionally, an initially placed coefficient may require changing at a later stage of the balancing process. For example, consider the stepwise balancing of the following equation.

$$Ag_2O \rightarrow Ag + O_2$$

$$Ag_2O \rightarrow 2\,Ag + O_2$$
(To initially balance Ag.)

$$2\,Ag_2O \rightarrow 2\,Ag + O_2$$
(To balance O, but Ag is now unbalanced.)

$$2\,Ag_2O \rightarrow 4\,Ag + O_2$$
(Change of coefficient to balance Ag. Entire equation now balanced.)

By convention, generally the smallest set of whole numbers is used for the coefficients in a final balanced chemical equation. It may be necessary either to divide all coefficients by the same number to yield a lower set of whole numbers or to multiply all coefficients by the same number to remove partial or fractional numbers. Note the following examples.

1 $$6\ Na_2CO_3 + 4\ H_3PO_4 \rightarrow 4\ Na_3PO_4 + 6\ CO_2 + 6\ H_2O$$
(A balanced equation.)

$$3\ Na_2CO_3 + 2\ H_3PO_4 \rightarrow 2\ Na_3PO_4 + 3\ CO_2 + 3\ H_2O$$
(Produced by dividing all
coefficients by 2. Still balanced.)

2 $$2\ Al + 1.5\ O_2 \rightarrow Al_2O_3$$
(A balanced equation.)

$$4\ Al + 3\ O_2 \rightarrow 2\ Al_2O_3$$
(Produced by multiplying
all coefficients by 2.
Still balanced.)

This chapter has provided a study of the various types of matter and some of their chemical characteristics. Three forms of chemical notation have been described — chemical symbols, formulas, and equations. It now would be appropriate to ask how chemical formulas for compounds can be predicted so that correct chemical equations can be written for chemical reactions. It also would be appropriate to inquire about predicting chemical behavior of elements and compounds, which will be the topic of discussion in the next chapter.

study questions

1 What distinguishes pure matter from impure matter?
2 List five examples of pure matter and five examples of impure matter.
3 How many elements are known presently? Are all these naturally occurring?
4 What are the four most abundant elements in the earth's crust?
5 The names of elements may be derived from several sources. What are some of these sources?
6 What are the chemical symbols for the following elements?
a. Iodine   c. Zinc   e. Neon
b. Calcium   d. Chromium f. Tungsten

7 What elements are represented by the following symbols?
a. Ni   c. Pb   e. Si
b. Sn   d. U   f. As

8 Is there a difference between CO and Co? Why?

9 What law permits writing discrete chemical formulas for each chemical compound?

10 Can the elements involved in the formation of a compound be separated and isolated by physical means? Why?

11 What elements are present in each of the following substances, and what is the combining ratio of atoms in each case?
a. HNO_3 b. K_3PO_4 c. $(NH_4)_2SO_4$

12 Decide whether the following are elements, compounds, or mixtures.
a. Water   d. Soil   g. Air
b. Sand   e. Wood   h. Oxygen
c. Aluminum   f. Sulfur

13 What one property distinguishes an ion from an atom or a molecule?

14 Label each of the following as physically homogeneous or heterogeneous and chemically homogeneous or heterogeneous.
a. Air   b. A gold coin   c. An ice cream soda

15 What are the benefits of using chemical symbols, formulas, and equations?

16 Balance the following chemical equations.
a. $C + O_2 \rightarrow CO_2$
b. $C_2H_6 + O_2 \rightarrow CO_2 + H_2O$

17 Why is it necessary to balance chemical equations?

18 Distinguish between a subscript and a coefficient in a chemical equation.

special problems

1 Bottled gas (propane) is a compound consisting of only carbon and hydrogen. If a sample of propane contains 36 grams of carbon and 8 grams of hydrogen, what is its percentage composition? How much carbon is found in 150 grams of propane?

2 Analysis of Epsom salts indicates that it consists only of magnesium and chlorine. If a 95-gram sample of Epsom salts contains 24 grams of magnesium, what is the percentage

composition of this compound? How much chlorine would be found in a pound of Epsom salts? Express your answer in grams.

3 Balance the following chemical equations. All chemical formulas are correct as written.

a. $C_2H_6 + O_2 \rightarrow CO_2 + H_2O$

b. $Mg + O_2 \rightarrow MgO$

c. $Na + H_2O \rightarrow NaOH + H_2$

d. $AgNO_3 + CaCl_2 \rightarrow Ca(NO_3)_2 + AgCl$

e. $Fe + O_2 \rightarrow Fe_2O_3$

chapter **4**

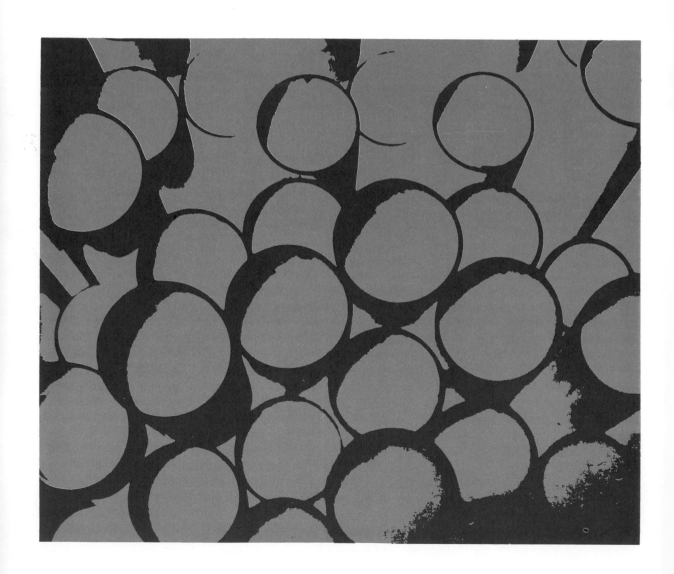

fundamental particles and nuclear structure

To understand how chemical compounds are formed and why chemical reactions take place, it is first necessary to understand the structure of atoms and molecules. Knowledge in these areas has evolved slowly over many centuries and is still incomplete. This chapter briefly discusses the background for the atomic theory and then discusses atomic structure with emphasis on the nucleus. The electronic structure of the atom is the topic of Chapter 6.

early ideas

Speculation on the structure of matter has gone on since the dawn of civilization. Democritus, a Greek philosopher of the fifth century B.C., usually is credited with being one of the first to propose an atomic theory. Democritus believed that there was a limit to the extent that matter can be divided. He felt that if a piece of matter were subdivided continuously, at some point an indivisible particle would be obtained. The term **atom** that is used for these particles today is derived from the Greek word *atomos* meaning indivisible.

About a century later, under the influence of Aristotle, the atomic theory of Democritus was discarded and replaced with a new theory. According to Aristotle, all matter consisted of four basic elements: fire, air, earth, and water. The theory also maintained that all other substances could be prepared by the proper combination of these basic elements.

The Aristotelian theory predicts, then, that it would be possible to convert some material of relatively low value, such as iron, into a material of relatively high value, such as gold, by the proper combination of basic elements. Thus the Aristotelian theory was undoubtedly responsible for the misguided attempts of the alchemists to make gold from the "base metals."

atom
electrode
cathode
anode
cathode ray
electron
proton
radioactive element
alpha ray
beta ray
gamma ray
atomic number
ion
atomic mass units
isotope
neutron
gram-atom
Avogadro's number

During the first decade of the nineteenth century, John Dalton, an English schoolteacher, proposed a new atomic theory, which modified and strengthened the atomic concept of matter. Basically, the theory incorporated the following ideas.

1 Elements are composed of indivisible particles called atoms.
2 Atoms can be neither created nor destroyed.
3 Atoms of the same element have the same mass and size, but atoms of different elements have different masses and sizes.
4 Chemical compounds are formed by the combination of two or more elements.
5 Atoms combine in simple numerical ratios, for example, 1:1 or 1:2, but never 1:$\frac{1}{2}$.

The two theories discussed here point out some of the characteristics of theories in general. A *theory* is merely a proposed explanation for an observed phenomenon. As such, it must fit all known facts that exist at the time it is proposed. *A theory is not necessarily correct.* Theories are used throughout science and are rarely preserved in unmodified form for long periods, because new facts emerge requiring that the original theory be modified or discarded completely. As the discussion of atomic theory progresses, it will be seen that this is a good example of a theory that has been extensively modified. The present ideas of atomic structure are a result of this process. These ideas may well be modified further as new facts emerge.

An interesting phenomenon frequently associated with theories is that the longer they last, the less willing people become to discard them. Sometimes the need for change is quite compelling; yet a well-ingrained, erroneous theory will be perpetuated, owing to an unwillingness to adapt to new situations and ideas. The Aristotelian theory lasted for about 2000 years before it became overwhelmingly obvious that the theory no longer fit the observations. The final blow came when air was shown to be a mixture of gases, not an element. This occurred in the early eighteenth century and led to a revival of the atomic theory.

Later, more sophisticated experiments have demonstrated that these ideas were only partially correct. The ideas fit experimental observations so well, however, that the model of the atom as a simple, hard sphere was accepted until the end of the nineteenth century. In fact, for most practical situations, these concepts can be successfully applied without modification even today.

Toward the end of the nineteenth century, new techniques were developed for studying particles of atomic size. For example, X rays were discovered in 1895, radioactivity in 1896, the electron in 1897, and radium in 1898. The hard-sphere model of the atom was inconsistent with these new discoveries. The new concepts that arose as a result form the basis of much modern understanding of atomic structure and chemical behavior.

electrical nature of matter

Electrical phenomena have impressed man for so long as he can remember. The fury of a lightning storm is awesome. Even the sparks generated by stroking an animal's fur on a dry, windy day are impressive. An understanding of these phenomena, however, was not developed until the nineteenth century. Prior to that time, superstition was invoked to explain them. The anger of the god, Zeus, for example, was the explanation offered by the ancient Greeks for lightning.

Ancient Greeks believed that lightning reflected the anger of the god Zeus. Today, however, it is known that lightning is a natural phenomenon created by the interaction of oppositely charged matter. (*Courtesy Westinghouse Electric Corporation*)

Early experiments on electrical phenomena involved rubbing objects like glass or hard rubber rods with silk or animal fur. Electrical charges thus produced were of two types, labeled *positive charges* and *negative charges*. It was observed that objects bearing like charges repel each other, but objects bearing unlike charges attract each other.

The phenomenon of static electricity is well known today. Almost everyone has experienced a static discharge after walking on a carpet made from synthetic fibers. The charge that can be built up in this manner is large enough to generate a spark over a centimeter long. A simple experiment that can be conducted easily is to run an ordinary plastic comb through your hair and then pick up small pieces of tissue paper by holding the comb near them. Simple experiments like this were the forerunners of the more sophisticated experiments that gave us our present understanding of the composition and structure of matter.

Discoveries concerning the composition and ultimate source of electrical charges were impossible until a ready source of electricity became available in 1800, when Volta produced the first electrochemical cell. The early voltaic cells were the forerunners of the modern dry cell used in flashlights and the lead storage battery of automobiles. (Operation of dry cells and lead storage batteries is discussed in Chapter 10.) Once these sources of electricity became known, progress was rapid. Not only was an understanding of electricity achieved, but also useful applications, such as electrical appliances, motors, radio, and television, were developed.

The experiments that led to an understanding of electricity and atomic structure were conducted by passing an electrical current through small glass tubes, from which most of the air had been removed. Evacuated tubes were used because electricity will flow through the relatively empty space between two metal plates placed within a tube. A similar charge will not pass through the space when air is present in a tube.

A simple tube of this type is shown in Figure 4-1. This tube contains two metal plates, called electrodes, that are connected to a source of electricity. The negatively charged electrode is called a cathode, and the positively charged electrode is called an anode.

When a small amount of gas is introduced into the tube, the tube will glow. The color varies with the nature of the gas introduced. Neon gas, for example, gives an orange-red glow; hydrogen is nearly white. Neon signs, a commercial adaptation of these tubes, are made by inserting electrodes into glass tubes that have been bent into the shape of letters and other designs. Colored signs are made by varying

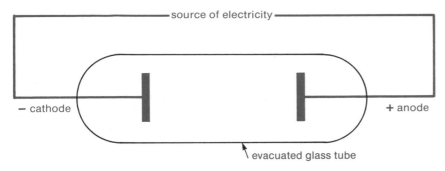

− cathode   + anode

evacuated glass tube

Figure 4-1.   A simple evacuated tube for studying the properties of electricity

The neon signs that brighten city streets are made of gas-filled tubes that glow when an electrical current is passed through them. (*Courtesy Greater Reno Chamber of Commerce*)

the gases in the tubes. The origin of the various colors will become apparent in Chapter 6, when the electron configuration of atoms is discussed.

Fluorescent lamps such as those used in most stores, factories, and schools are similar to neon signs. The gas usually contains a mixture of mercury vapor and krypton. When an electrical current passes through the fluorescent tube, the gases emit light, primarily in the ultraviolet or invisible region of the spectrum. A solid material known as a *phosphor* is coated on the inside surface of the tube, absorbing the ultraviolet light and re-emitting it as visible light.

---

cathode rays

A number of experiments have been devised to show that the "rays" causing the light in the tube travel from the cathode to the anode. For this reason, these rays are called cathode rays. One such experiment involves the tube in Figure 4-2. When the electrodes are arranged as shown, the anode casts a sharply defined shadow on the end of the tube. When the charges on the electrodes are reversed, no shadow is observed.

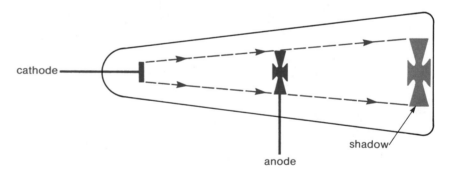

Figure 4-2. A discharge tube designed to demonstrate the existence of cathode rays

---

Another type of tube used in the study of cathode rays is shown in Figure 4-3. In this tube, a metal plate containing a slit is placed between the anode and the cathode to stop all but a narrow beam of cathode rays. A zinc sulfide screen is placed nearly parallel to the path of the beam in such a way that most of the beam comes in contact with the screen. Light is emitted wherever the cathode rays

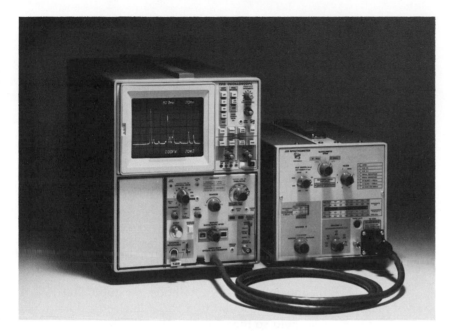

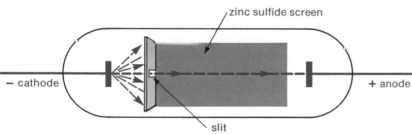

Figure 4-3.   A cathode-ray tube designed to study the magnetic properties of cathode rays

strike the zinc sulfide. Under normal circumstances, the beam travels in a straight line between the cathode and the anode. The beam is deflected, however, by a magnetic or electrical field. The cathode-ray beam is attracted toward a positive field and away from a negative field. Since like charges repel each other and unlike charges attract, the cathode ray must be negatively charged. This experiment demon-

Cathode-ray tubes are used in many modern instruments and devices. These include oscilloscopes, such as those that monitor heartbeats or are used in electronics, TV picture tubes, in which the electron gun in the back or narrow end of the tube is the cathode, and radar scopes.

strates that cathode rays are particles of matter. Matter is affected by a magnetic field, but energy is not. Light rays, for example, are not bent by a magnetic field.

These and other studies have shown that cathode rays have the following properties.

1 Cathode rays travel in a straight line away from the cathode.
2 Light is emitted when cathode rays strike zinc sulfide.
3 Cathode rays are composed of particles of negatively charged matter.
4 The path of cathode rays is bent by a magnetic or electrical field.
5 The nature of cathode rays is independent of:
 a. The nature of the cathode.
 b. The residual gas in the tube.
 c. The kind of wire used to conduct the current to the tube.
 d. The source of electricity.

In 1897, J. J. Thomson devised a cathode-ray tube that had a known magnetic field perpendicular to a variable electrical field. He adjusted the electrical field so that the deflection of the beam caused by the magnetic field was balanced exactly by the electrical field. By measuring the electrical field, Thomson determined the ratio of charge to mass for the newly discovered particle, called an electron.

The charge on the electron was not determined until 1909, when R. A. Millikan performed the classical oil-drop experiment, illustrated in Figure 4-4. In this experiment, Millikan used an atomizer

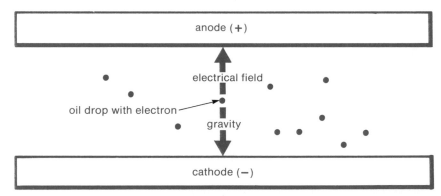

Figure 4-4. Millikan's oil-drop experiment

similar to that found on perfume bottles to spray oil droplets into a chamber containing charged metal plates on the top and bottom. If he were to conduct this experiment today, he probably would use an aerosol spray to generate the oil droplets. Electrons were released from atoms in the air by irradiating the chamber with X rays. Normally, the drops would fall to the bottom because of gravity. If the droplet of oil picked up an electron, however, it would be attracted upward to the positive plate. By balancing the electrical field with the gravitational attraction, it was possible to suspend the drop. By measuring the electrical field required to achieve this condition, the charge on the particle could be determined. Of course the charge varied from drop to drop, because some drops picked up more than one electron. A minimal charge was found, however, from which the other charges could be obtained by multiplying that charge by a whole number. This minimal charge was assumed to be the charge on one electron.

Based on these experiments and several others, the electron is now assigned a charge of -1 and a mass $1/1838$ times that of the hydrogen atom.

positive rays

With the discovery of electrons, scientists immediately began looking for a positively charged particle. Such a particle now was expected, because matter is electrically neutral under ordinary conditions. The key initial experiment that led to the discovery of the positive particle was conducted by Goldstein in 1886, actually before the discovery of the electron.

In his studies, Goldstein had been using a perforated cathode and had observed a glow on the back side of the cathode. Later workers obtained well-defined positive rays, using tubes similar to the cathode-ray tubes already described. These positive rays were studied by techniques similar to those used to study cathode rays. The positive rays were found to have a mass-to-charge ratio that varied with the nature of the gas in the tube. With hydrogen gas, a particle was found that had a positive charge equal in magnitude to that of the electron. This particle, which was the smallest one found, had a mass about $1837/1838$ times that of a hydrogen atom. This smallest particle with unit positive charge now is called a **proton** (Figure 4-5).

natural radioactivity

In 1896, Henri Becquerel discovered that, when placed near uranium, an unexposed photographic plate became exposed. This was true

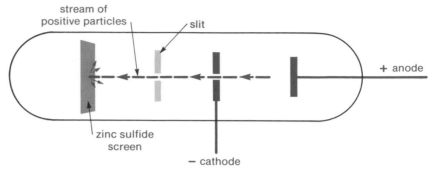

stream of
positive particles

slit

+ anode

zinc sulfide
screen

− cathode

Figure 4-5. Apparatus used in the discovery of the proton

even when the plate had not been removed from its sealed paper container. Thus it was found that uranium was emitting radiation that could pass through paper. Shortly thereafter, Marie Curie discovered the element radium, which also had this property. Elements that emit radiation more penetrating than light soon became known as radioactive elements.

The Daltonian idea that atoms were indestructible had been disproved thoroughly. First, charged particles were found to be produced from atoms. Now, suddenly, naturally occurring elements were discovered that spontaneously decomposed to form other elements. Obviously, atoms were not indestructible.

When radioactive elements such as radium were placed in a magnetic or electrical field, an interesting discovery was made. It was

Pitchblende, or uranitite. This ore is mined principally for its uranium content. It was from pitchblende that the Curies isolated radium for the first time. (*Courtesy Westcott, U.S. Atomic Energy Commission*)

Marie Sklodowska Curie was born in Warsaw, Poland, in 1867. She wished to study medicine, but received degrees in physics and mathematics instead. She and her husband Pierre, also a physicist, became intrigued by the phenomenon of radioactivity and pursued investigations to learn more about its properties. After four years of extremely difficult work, these investigations resulted in the isolation from pitchblende of radium and polonium (named after Poland). Although radioactivity was discovered by Becquerel, it was actually Marie Curie who gave the phenomenon its name. For their discoveries, the Curies were awarded the Nobel Prize in physics in 1903. In 1911, Marie Curie was given a rare honor by being awarded a second Nobel Prize, this time in chemistry, for her further work with radium. She died in 1934, from an incurable blood disease that resulted from overexposure to radium.

observed that three different types of rays could be emitted. These rays are called alpha (α), beta (β), and gamma (γ). The effect of a magnetic field on these rays is shown in Figure 4-6. The α rays appeared to be positively charged. Later work demonstrated that the α rays consist of doubly positively charged helium nuclei. The β rays were negatively charged and later were found to consist of electrons. The γ rays were not deflected by the magnetic field; thus they had no charge. They were assumed to be pure energy and without mass. Radioactivity will be discussed in greater detail in Chapter 5.

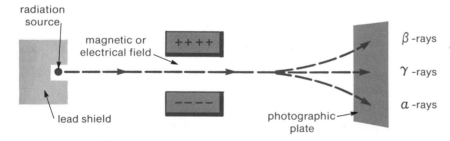

Figure 4-6.   The effect of an electrical or magnetic field on the three types of radioactive rays

nucleus of the atom

Ernest Rutherford, an English physicist, became interested in α particles. In 1909, he reported one of the classical experiments of this era. Rutherford bombarded a thin gold target with α-particles in a tube similar to that shown in Figure 4-7.

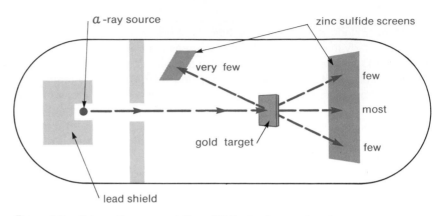

Figure 4-7. Schematic representation of Rutherford's experiment

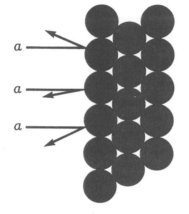

Figure 4-8. Expected deflections of alpha particles by solid atoms

All previously proposed models represented atoms as solid particles. Thomson, for example, pictured the atom as a positively charged sphere with the negatively charged electrons dispersed at random within it. If such a model were correct, all α particles should have been reflected, as shown in Figure 4-8. This was not Rutherford's observation. He observed instead that most α particles went straight through the target unchanged in direction of travel. A few particles were deflected partially, but very few actually were reflected back toward the α particle source.

As is often the case when new, unexpected observations are made, it took considerable time to develop an adequate explanation for these experimental results. It was not until 1911 that Rutherford had an explanation consistent with the observation. Like most such explanations, it seems simple and obvious in retrospect.

Figure 4-9. Rutherford's model of the atom explaining alpha-particle behavior

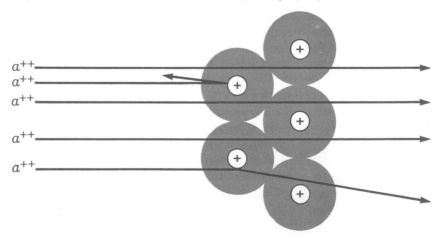

Rutherford visualized an atom as having a very small, heavy nucleus. This nucleus must have a positive charge, since it repels α particles, and it must be much heavier and more highly charged than an α particle, since some of the α particles were reflected. Since an atom is electrically neutral, its bulk must be composed of empty space and electrons. Thus it would be possible for the α particles to pass through the atom, as shown in Figure 4-9.

More recent investigations indicate that an atom has a diameter of about 10^{-8} centimeters, and the nucleus has a diameter of about 10^{-13} centimeters. In more familiar terms, if a grain of sand represents an average nucleus, then the outer edge of the atom would be the length of a football field away from this grain of sand.

atomic number

At about the same time that Rutherford was developing his new atomic model, H. G. J. Moseley was working with X rays. X rays are produced when high-energy electrons strike a metal target. A simple X-ray tube is shown in Figure 4-10.

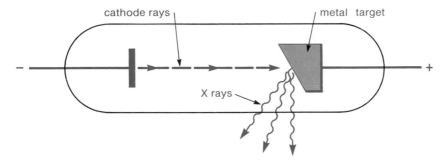

Figure 4-10. A simple X-ray tube

Moseley discovered that X rays of a characteristic wavelength are produced for each type of metal target used. He further observed that, with only a few exceptions, there was a regular change in wavelength with increasing atomic weight. Lighter elements gave the longest wavelength X rays; heavier elements gave shorter wavelength X rays.

Moseley concluded that the number of positive charges on the nucleus increases from element to element by a single electrical charge. This increase is related to the number of protons in the nucleus. The number of protons in the nucleus, and therefore the number of electrons in a neutral atom, is now called the **atomic number.** Each element has its own, unique atomic number. Most systematic arrangements of elements, called periodic tables, show the atomic number of each element. The atomic number increases progressively throughout the table as shown in Figure 4-11.

Several periodic tables had been proposed prior to Moseley's time. One of the more significant, originated by Mendeleev, arranged the elements according to their chemical and physical properties. The new arrangement according to atomic number correlated well with Mendeleev's arrangement. The arrangement of elements shown in Figure 4-11 will be discussed in more detail after electron configurations of the elements have been described.

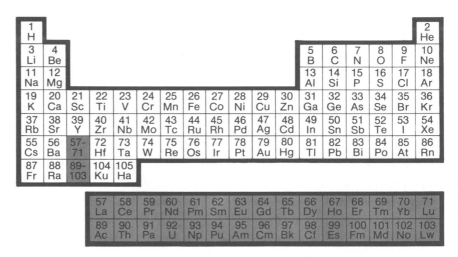

Figure 4-11. A simplified periodic table showing the atomic number for each element

discovery of isotopes

Around World War I, some investigators became involved with an extension of the cathode-ray tube concept that eventually led to the mass spectrograph. A simplified drawing of this instrument is shown in Figure 4-12.

In this mass spectrograph, a sample is vaporized and introduced into the sample chamber, where an electrical discharge takes place. The electrons in the discharge dislodge electrons from the sample so that the sample becomes positively charged or ionized. An **ion** is an

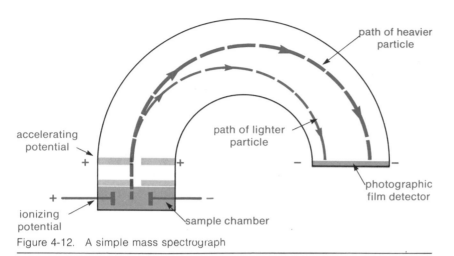

accelerating
potential

path of heavier
particle

path of lighter
particle

photographic
film detector

ionizing
potential

sample chamber

Figure 4-12.   A simple mass spectrograph

A mass spectrograph. Use of the mass spectrograph led to the discovery of isotopes. (*Courtesy Nuclide Corporation*)

atom or group of atoms containing more or less electrons than its total proton number requires; therefore it is electrically charged. The ions are accelerated toward a cathode, which has a slit so that a beam of these particles passes through the electrode and into the influence of a strong magnetic field. The magnetic field is placed so that the charged particles follow a curved path. The radius of curvature depends on the magnetic field strength and the mass of the charged particles. A particle with lighter mass would follow a path with a smaller radius of curvature. If photographic film is placed at the far end of the instrument, the charged particles will affect the film wherever they strike it. The resulting mass spectrum will resemble Figure 4-13.

The position of the band on the film is related to the mass of a particle. If the magnetic field strength is correlated with atomic mass units, then the atomic weight of each particle can be obtained directly. Atomic mass units (amu) are relative mass units based on the mass of a carbon atom that is assigned the value of 12. A further discussion of this concept is given in the next section.

As a result of studies using the mass spectrograph, it was discovered that all atoms of a given element do not necessarily have the same weight. Most neon atoms, for example, have an atomic weight of 20 amu, but some atoms weigh 22 amu. Magnesium has three

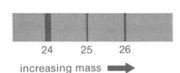

24   25   26

increasing mass ➡

Figure 4-13. A mass spectrum of magnesium

Since their discovery, neutrons have been used in various ways. For example, radiation produced by bombardment with a high-intensity beam of neutrons revealed an apparent clay filling in the leg of this 2000-year-old Chinese bronze ceremonial urn. (*Courtesy Tony Dapello, Atomics International*)

types of atoms, with masses of 24, 25, and 26 amu. Most magnesium atoms (78.8%) weigh 24 amu, only 10.1% weigh 25 amu, and 11.1% weigh 26 amu. Atoms of the same element that differ in mass are called isotopes.

Prior to discovery of isotopes, atomic mass was thought to be related solely to proton or atomic number. It was found, however, that the atomic weights of most elements actually exceeded the predicted weights by a large margin. Lithium, for example, with an atomic number of 3, has an atomic weight of 6.94 amu.

To account for these observations, a new particle was proposed, having the mass of a proton, but no electrical charge. If such a particle existed, it necessarily would be within the nucleus, and it could easily explain the observations described. Protons and electrons still would control the chemical properties of the atom, but the weight would result from both the protons and the new neutral particles. Isotopes then could be explained by assuming that the differences in weight were due to some atoms containing more of these neutral particles than other atoms of the same element. This new particle, called a neutron, finally was detected and confirmed by James Chadwick in 1932.

atomic weights

It is now possible to discuss the weight of atoms. Each atom has its own mass, which is approximately equal to the sum of the masses of the protons and neutrons in the nucleus. Although electrons are also present, they are too small to contribute significantly to the atom's mass. All atoms of the same element have the same number of protons, but they may have different numbers of neutrons, leading to different isotopes. This concept is illustrated in Table 4-1 with isotopes of magnesium.

This concept is satisfactory from a theoretical point of view, but it does not help much with the problem of actually determining the mass of an atom. Atoms are still too small to be weighed by any conventional technique. Nevertheless, it is possible to devise a scale

Table 4-1. Nuclear composition of isotopes of magnesium

Isotope Mass	Proton Number	Neutron Number
24	12	12
25	12	13
26	12	14

of relative atomic weights. The most obvious procedure would be to pick the lightest element, hydrogen, and give it a relative weight of one. Other elements could then be compared with hydrogen on this basis. For ease of calculation, however, carbon with a mass of 12 has been chosen as the reference. In other words, the particular isotope of carbon that contains six protons and six neutrons is the standard and is assigned the relative atomic weight of 12 atomic mass units. The periodic table (Figure 7-1) and the list of elements (Table 3-1) both give relative atomic weights based on this standard.

Since the mass of each element is due almost entirely to protons and neutrons, it would be logical to expect the atomic weights of the elements to be integers. This is not observed. The reason is that the atomic weights shown in the table are experimentally determined atomic weights for the naturally occurring elements. Naturally occurring elements consist of mixtures of the various isotopes of the elements. Therefore, the atomic weights that have been determined represent relative average weights of atoms of the elements.

Techniques for determining atomic weights have become so precise that differences in isotope ratio from sample to sample can be detected. A complication is now arising because of man's intentional separation of isotopes. Uranium, for example, is used extensively for atomic energy. The uranium isotope with mass number 235 is most valuable for this purpose. The more common isotope, with mass 238, is separated and used for industrial purposes. Uranium that has been depleted of its uranium-235 content no longer has the natural atomic weight. For precise work, the origin of the sample must be known.

the gram-atom and avogadro's number

The atomic weight as previously described is a relative atomic weight expressed in atomic mass units. For practical purposes, it is desirable to be able to use the atomic weight scale with conventional mass units. This is accomplished by using a unit known as the gram-atom. A gram-atom is merely the atomic weight of an element expressed in grams. For example, a gram-atom of hydrogen weighs 1.008 grams. If one gram-atom of each of two elements is measured, the same number of atoms of each element will be present in these samples. For example, 1.008 grams of hydrogen will contain the same number of atoms as 15.999 grams of oxygen.

Numerous ways have been developed to determine how many atoms actually are contained in a gram-atom of an element. There is remarkably good agreement that there are 6.02×10^{23} atoms in a gram-atom of any element. This figure is known as Avogadro's number in honor of Amedeo Avogadro, an Italian physics professor who did fundamental work in this area.

The actual weight of an atom cannot be measured directly. It can be estimated, however, by dividing the weight of a gram-atom of that element by Avogadro's number. The weight of a hydrogen atom, for example, can be calculated as follows.

$$\frac{1.008 \text{ g/g.atom}}{6.02 \times 10^{23} \text{ atoms/g.atom}} = 1.7 \times 10^{-24} \text{ g/atom}$$

other nuclear particles

In the past thirty or forty years, great strides have been made in nuclear chemistry and nuclear physics. New particles and energy forms, such as positrons, mesons, and neutrinos, have been proposed and ultimately discovered. Still more have been proposed, but have not been found yet. Today, for example, atomic physicists are looking for something they call a *quark*. Quarks are supposed to be fundamental particles of matter with a charge of one-third that of the electron. These exotic particles do not contribute to ordinary chemistry and may not even exist as such in the nucleus.

Changes in atomic structure that are significant in chemical reactions will be described in the next few chapters. The discussion will begin with a look at reactions of the nucleus. This discussion will be followed by a study of the electron arrangements in atoms and their influence on the chemical behavior of atoms.

Aerial view of the meson facility at the Los Alamos Scientific Laboratory in New Mexico. In the center is the long linear tunnel where particles are accelerated to high speeds. Expected uses of mesons include control and treatment of cancer. (*Courtesy Los Alamos Scientific Laboratory*)

study questions

1 What is the origin of the word *atom*?
2 What was the Aristotelian theory of matter?
3 What is an electrochemical cell? Who is credited with producing the first cell of this type?
4 List some properties of cathode rays.
5 What experimental observations showed that cathode rays were composed of negatively charged particles of matter?
6 What information about electrons was obtained by Millikan through his oil-drop experiment?
7 What experimental evidence disproved the idea that atoms were solid particles of matter?
8 Describe the structure of an atom.
9 What contributions to the development of the atomic theory of matter were made by the following individuals?
   a. Democritus   d. Bohr
   b. Dalton   e. Thomson
   c. Rutherford
10 Henri Becquerel often is referred to as the discoverer of radioactivity. What were the circumstances under which his discovery was made?
11 Why are α and β rays thought to be particles of matter, but γ rays are not?
12 Why may atoms of the same element have different atomic masses?
13 Of what value was the mass spectrograph to the understanding of the atomic makeup of elements?
14 Why must the origin of a sample be known for precise work involving atomic weights?

special problems

1 How many grams make up one gram-atom of the following elements?
   a. Oxygen   b. Silver   c. Barium
2 How many grams make up two gram-atoms of carbon?
3 How many grams are in 0.36 gram-atom of iron?
4 How many gram-atoms are in 24 grams of sulfur?
5 How many gram-atoms are in 157 grams of zinc?
6 How many electrons, protons, and neutrons are found in the following elements?

	Electrons	Protons	Neutrons	Mass
Magnesium (Mg)	12	_____	_____	24 amu
Aluminum (Al)	_____	13	14	_____
Rubidium (Rb)	37	_____	50	_____
Uranium (U)	_____	92	_____	238 amu
Carbon (C)	_____	_____	_____	14 amu

7 What is the weight in grams of a single atom of magnesium?

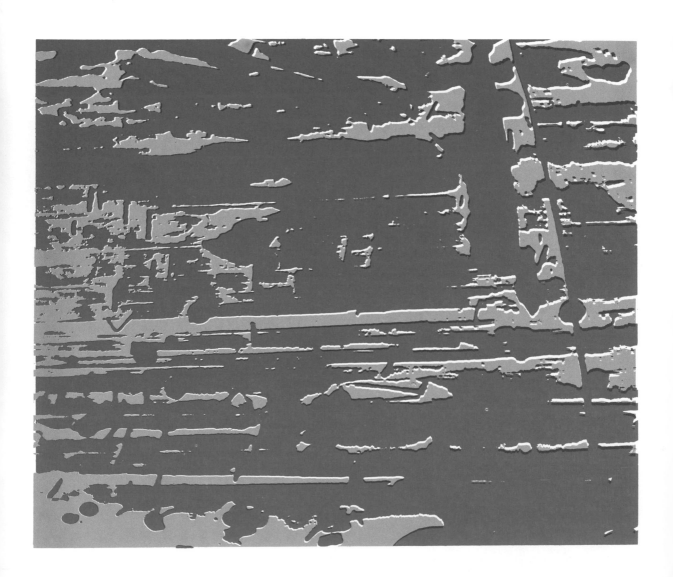

nuclear chemistry

When an element becomes involved in a chemical reaction, its atoms undergo changes in composition. Most reactions that an atom undergoes involve changes in its electronic structure, but not its nuclear structure. These common reactions, which will be described in greater detail in Chapter 8, lead to chemical compounds. An atom, however, may participate in a reaction in which its nuclear composition is altered. In reactions of this type, an element is converted into a different element, because its proton number almost always is changed. These nuclear reactions can occur spontaneously, or they may be induced artificially by man. This chapter will explore the characteristics and uses of nuclear reactions and atomic energy.

natural or spontaneous nuclear reactions

Most naturally occurring elements exist as mixtures of isotopes, variations of atoms of an element that differ in neutron number and hence mass. About 40 of these isotopes are unstable and spontaneously disintegrate or decay to form different elements. All isotopes of elements with atomic numbers of 84 and higher have this property. The disintegration process produces radiation of three major types, alpha (α), beta (β), and gamma (γ). An introductory discussion of these forms of radiation was provided in Chapter 4. Radiation from these isotopes often can be detected by instruments such as Geiger-Müller counters, which produce audible signals, and this nuclear phenomenon is labeled appropriately *radioactivity*. Historically, the term was introduced by Marie Curie, the discoverer of radium, in 1898, two years after the discovery of the phenomenon by Henri Becquerel.

Some of the principal characteristics of the three forms of radiation are summarized in Table 5-1. Gamma radiation travels at about

Table 5-1. Some properties of the major types of radiation from spontaneous nuclear reactions

Type	Electrical Charge	Relative Mass	Composition	Symbol or Designation
alpha	positive	4 amu	helium ions	$^{4}_{2}He$ or $^{4}_{2}\alpha$
beta	negative	$\frac{1}{1838}$ amu	high-energy electrons	$^{0}_{-1}e$ or $^{0}_{-1}\beta$
gamma	neutral	0 amu	energy waves	γ

Left: Portable Geiger counter used to measure radioactivity. *Right*: Laboratory-model scintillation counter used in laboratory measurements of radioactivity. (*Courtesy left, Eon Nuclear and Medical Corporation; right, Beckman Instruments, Inc.*)

the speed of light. Made of nonparticulate, high-energy waves, this form of radiation has the greatest penetrating power into solid matter and is mainly responsible for the burns obtained when one is exposed to nuclear radiation. Beta particles travel at about 90–95% the speed of light and can penetrate several centimeters into solid matter. Alpha particles, the heaviest of the three forms of radiation, travel most slowly (about 5–10% the speed of light) and can be stopped by a sheet of paper or human skin. When a particular atom undergoes a spontaneous nuclear reaction, the radiation that is emitted usually consists of alpha and gamma, beta and gamma, or sometimes only gamma radiation. No observations of simultaneous release of alpha and beta particles by a nucleus have yet been made experimentally.

The designations used for the alpha and beta particles, for example $^{4}_{2}He$, identify the composition of the particle (by symbol), its atomic number (by subscript), and its mass (by superscript.)

Since beta particles do not have atomic numbers, the subscript -1 on the left of the beta-particle symbol indicates its charge. This type of designation also will identify a specific isotope of an element in subsequent discussions. Since radioactivity is associated with only certain isotopes, designation of specific isotopes in discussions of nuclear reactions is important.

mass ⟶ $^4_2$He ⟵ element
atomic number ⟶ or particle

(or charge, for beta particles)

alpha emission

The loss of an alpha particle (alpha emission) from a nucleus results in the removal of four mass units and a decrease of two in the atomic number (or proton number), thus producing a new element. For example, radium-226 is a known alpha emitter, and its spontaneous disintegration may be represented by the following equation.

$$^{226}_{88}\text{Ra} \rightarrow \, ^4_2\text{He} + \, ^{222}_{86}\text{Rn} + \gamma \text{ radiation}$$

The product of this nuclear reaction is an isotope of radon with a mass number of 222. It is no longer radium because the atomic number also has changed.

Other examples of alpha emissions are shown below.

$$^{238}_{92}\text{U} \rightarrow \, ^4_2\text{He} + \, ^{234}_{90}\text{Th} + \gamma \text{ radiation}$$

$$^{210}_{84}\text{Po} \rightarrow \, ^4_2\text{He} + \, ^{206}_{82}\text{Pb} + \gamma \text{ radiation}$$

In nuclear reactions such as those shown, sums of masses and charges or atomic numbers are equal on both sides of the arrow. The balance of mass and charge is identical in concept to that used in writing correct equations for the more common chemical reactions.

beta emission

As in alpha emission, a different element results when a nucleus undergoes beta emission. For example, when $^{60}_{27}$Co undergoes beta emission, $^{60}_{28}$Ni is produced.

$$^{60}_{27}\text{Co} \rightarrow \, ^0_{-1}e + \, ^{60}_{28}\text{Ni} + \gamma \text{ radiation}$$

To understand this type of reaction, several characteristics of beta emissions require further explanation. First, the beta particle, which has been identified as a high-energy electron, originates from the nucleus and is not one of the extranuclear electrons found in an atom. Second, the emission of a beta particle results in an increase of the proton number by one, with no increase in mass. These phenomenena currently are explained by the theory that a neutron may be regarded as resembling a combination of a proton and an electron. When beta emission occurs, a neutron is believed to split into an electron, which is emitted as the beta particle, and a proton, which is retained in the nucleus. Since a neutron and a proton are nearly identical in mass, no significant change in mass is observed.

$$\begin{array}{ccc} {}^{1}_{0}n \rightarrow & {}^{1}_{1}p & + & {}^{0}_{-1}e \\ \text{neutron} & \text{proton} & & \text{electron} \\ & \text{(retained)} & & \text{(emitted)} \end{array}$$

Other examples of beta emissions follow.

$$^{27}_{12}Mg \rightarrow {}^{0}_{-1}e + {}^{27}_{13}Al + \gamma \text{ radiation}$$

$$^{214}_{82}Pb \rightarrow {}^{0}_{-1}e + {}^{124}_{83}Bi + \gamma \text{ radiation}$$

half-lives of isotopes

Radioactive isotopes have varying degrees of instability. The rate at which an isotope undergoes nuclear disintegration is represented by its half-life. The half-life of an isotope is the time required for one-half of a given sample of radioactive material to undergo nuclear disintegration. If, for example, one gram of an original two-gram sample of isotope A were observed to undergo nuclear decay in a period of three days, the half-life of A would be three days. Furthermore, within the next three-day period, one-half of the remaining one gram of isotope A would disintegrate, leaving only 0.5 gram of unreacted isotope. The amount of unchanged radioactive material is therefore halved in each half-life period.

Half-lives can be as short as a few millionths of a second or as long as several billion years. They do not depend on temperature and are virtually unalterable by man. The half-lives of some isotopes are given in Table 5-2.

Knowledge of the half-life of an isotope is of interest and importance to a scientist, because it enables him to know how long a particular isotope sample with which he is working will be active and present. Through the half-lives of carbon-14, uranium-238, or potassium-40, the approximate dates of archaeological remains and the approximate ages of the earth and rocks can be determined.

Table 5-2. Half-lives of some radioactive isotopes

Isotope	Half-Life
$^{238}_{92}U$	4.5×10^9 years
$^{40}_{19}K$	1.3×10^9 years
$^{14}_{6}C$	5,730 years
$^{60}_{27}Co$	5.27 years
$^{32}_{15}P$	14.3 days
$^{218}_{84}Po$	3.05 minutes
$^{12}_{5}B$	2×10^{-2} seconds

carbon-14 dating

The half-life of the radioactive isotope of carbon, carbon-14, is 5730 years. This isotope is a beta emitter. Under ordinary conditions in living matter, the ratio of carbon-14 atoms to those of the total non-radioactive carbon-12 atoms is about 1 to 10^{12}. This ratio is maintained through replacement and incorporation of carbon-14 from the atmosphere. Carbon-14 in the atmosphere is formed continuously by cosmic-ray interaction with nitrogen atoms.

When living matter dies, radioactive carbon is lost slowly through natural nuclear decay and is not replaced. Consequently, by determining the ratio of carbon-14 to the total nonradioactive carbon content in an archaeological specimen, comparing this with the ratio of these atoms in living matter, and using the half-life of carbon-14, a scientist can date the specimen approximately. The carbon-14 dating process was developed by Willard Libby at the University of Chicago in 1947. It can be used for specimens less than about 50,000 years old, and provides results generally accurate to within 600 or 700 years.

Radioactive carbon dating has been a valuable tool for studying life on this planet, including the early history of man and ancient civilizations. Dating can be done on various carbon-containing materials, including wood, charcoal, seashells, and bones. Through radioactive dating, the first appearance of man on the North American continent has been placed at about 11,000 years ago. Plant fossils found with Folsom-man artifacts near Clovis, New Mexico, show these remains are some 10,500 years old. Wood found with mastodon remains at Kings Ferry, New York, is also about the same age. In comparison, a tree from a tar pit in Los Angeles, California (from which Pleistocene mammals have also been removed) was dated by carbon-14 and found to be about 15,400 years old.

These human bones, uncovered in the Aswan Reservoir area in Sudan, were found to be 14,000 years old by carbon-14 dating measurements. (*Courtesy Anthropology Department, Southern Methodist University*)

geological dating

The product of a spontaneous nuclear disintegration is generally more stable than the starting isotope, but it may itself be radio-active. It is possible to have disintegrations connected in a series of reactions that ultimately leads to a stable isotope. In nature, three such series have been identified. One starts with uranium-235 and terminates with lead-207 in 11 reactions. Another starts with uranium-238 and terminates with lead-206 in 14 steps. The third starts with thorium-232 and terminates with lead-208 in 10 steps.

$$
\left.
\begin{array}{l}
{}^{235}_{92}\text{U} \xrightarrow{\text{11 steps}} {}^{207}_{82}\text{Pb} \\[1.5em]
{}^{238}_{92}\text{U} \xrightarrow{\text{14 steps}} {}^{206}_{82}\text{Pb} \\[1.5em]
{}^{232}_{90}\text{Th} \xrightarrow{\text{10 steps}} {}^{208}_{82}\text{Pb}
\end{array}
\right\} \text{stable isotopes}
$$

Because of the stability of the lead isotopes with which these natural series end, it has been possible to use half-life data to calculate the age of rocks or the age of the earth itself. For example, the method might involve the measurement of the U-238/Pb-206 ratio in a geological sample, followed by the calculation of the time required for the natural disintegration processes to produce this ratio of isotopes. The half-lives of members that make up the series involved would be needed to complete the calculations. Determinations of this type have revealed the age of the earth as about 4.5 billion years. The method is suitable for specimens that are more than 300 million years old.

Because potassium-containing minerals are more widespread in nature than uranium-containing ones, another dating method has been developed. Potassium-40 is a radioactive isotope that principally (88%) disintegrates to calcium-40. The remaining 12% of potassium-40 atoms disintegrates to argon-40. By determining the potassium-40/argon-40 ratio in a sample and using the half-life of potassium-40 (1.3×10^9 years), scientists can determine the age of that sample. This method has identified a granite deposit in the state of Washington as 17 million years old, plus or minus 0.5 million years, placing its formation in the Miocene period; a granite deposit in Georgia has been analyzed similarly and found to be 165 million years old, plus or minus 3 million years. More recently, this technique has been applied to rocks brought back from the moon by the Apollo astronauts.

It is generally believed that the moon was formed about the same time as the earth, or about 4.5×10^9 years ago. Because of volcanic activity as well as meteorological bombardments, however, the moon's surface has undergone significant chemical changes, and no primitive or original lunar material has been obtained for study thus far. The radioactive dating samples brought back by the Apollo astronauts indicate that the lunar crust material found on the highlands was formed about 4.1×10^9 years ago. In comparison, the material in the lunar "seas," or broad depressions on the moon's surface, has a different origin from the crust material and appears to have been formed by volcanic action. The igneous rocks (basalt) from the Ocean of Storms and Hadley's Rille show ages of about 3.3×10^9 years; those from the Sea of Tranquillity and the Sea of Serenity are older, dating back about 3.7×10^9 years.

The potassium-40/calcium-40 ratio is not used for calculating the age of geological samples because calcium-40 is so widespread in nature that the origins of most samples of this isotope are not radioactive disintegrations of potassium-40.

Moon rock sample, weighing 4.515 kilograms, brought back by the Apollo 15 team. Potassium-40/argon-40 measurements of some lunar samples have given information about the possible age of parts of the moon. (*Courtesy NASA*)

induced or artificial nuclear reactions

A stable nucleus that does not show any tendency for spontaneous decay may be forced to undergo a reaction by bombarding it with high-speed particles. These induced or artificial nuclear processes generally convert the stable nucleus into a radioactive one. It is through this means that certain radioactive isotopes that do not occur naturally can be prepared for scientific, medical, or commercial use, and it is also through this process that new synthetic elements have been prepared.

The first report of a man-induced nuclear reaction was made by Ernest Rutherford in 1919. He bombarded a sample of nitrogen with alpha particles and observed that hydrogen was formed. This observation led Rutherford to the conclusion, based on mass and charge balances, that the other conversion product from the nitrogen was oxygen-17. The following equation summarizes the reaction that occurred.

$$^{14}_{7}N + ^{4}_{2}He \rightarrow ^{1}_{1}H + ^{17}_{8}O$$

It is likely that a very unstable intermediate nucleus, called a compound nucleus ($^{18}_{9}F$ in this case), initially is formed by the bombardment and rapidly disintegrates by loss of a proton to form the atom of oxygen-17.

Although Rutherford conducted the first known artificial nuclear reaction, the product formed was not radioactive. The first reaction in which radioactivity actually was induced was achieved in 1934 by Irene Joliot-Curie, the daughter of Madame and Pierre Curie. She and her husband observed that, upon bombardment with alpha particles, aluminum-27 produced neutrons and phosphorus-30. Phosphorus-30, which has a half-life of about 2.5 minutes, is a radioactive isotope not previously known to exist in nature. It undergoes nuclear decay to produce silicon-30 and particles called positrons, which are positive electrons.

$$^{27}_{13}Al + ^{4}_{2}He \rightarrow ^{30}_{15}P + ^{1}_{0}n$$

$$^{30}_{15}P \rightarrow ^{30}_{14}Si + ^{0}_{+1}e$$

Positron emission is a less common form of radiation than alpha, beta, or gamma radiation.

Since these significant "first" instances of induced nuclear reactions, many others have been accomplished and reported. Bombardment particles now include protons ($^{1}_{1}H$), neutrons ($^{1}_{0}n$), deuterons ($^{2}_{1}H$), electrons ($^{0}_{-1}e$), and even ions of elements of low atomic number, such as $^{16}_{8}O$ and $^{12}_{6}C$. Additional examples of induced nuclear reactions are represented by the following equations.

$$^{59}_{27}Co + ^{1}_{0}n \rightarrow ^{60}_{27}Co$$

$$^{31}_{15}P + ^{2}_{1}H \rightarrow ^{32}_{15}P + ^{1}_{1}H$$

$$^{23}_{11}Na + ^{1}_{1}H \rightarrow ^{23}_{12}Mg + ^{1}_{0}n$$

For the bombarding particles to be effective in these reactions, they must collide with the target atoms at very high speeds. Modern technology has developed devices that serve as sources of high-speed particles or that accelerate particles to the desired velocity. Cyclotrons, linear accelerators, van de Graaf electrostatic generators, and synchrotrons are used to accelerate positively charged particles. Betatrons accelerate negatively charged particles, and nuclear reactors serve as principal sources of high-speed neutrons. The charged-particle accelerators increase the velocity of projectile particles by attracting them with oppositely charged fields along a circular or

Cyclotron at the Brookhaven National Laboratory in Upton, Long Island. Positive ions accelerated in a cyclotron are allowed to bombard stable atoms and convert them into radioactive isotopes. (*Courtesy Brookhaven National Laboratory*)

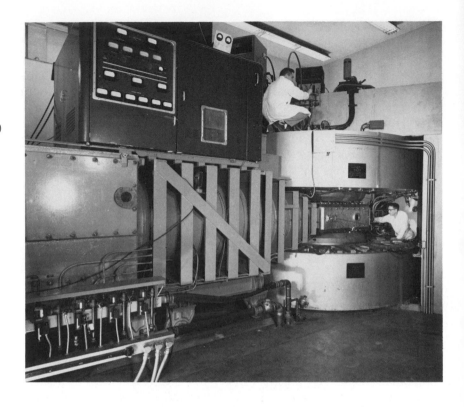

linear path. In nuclear reactors, uranium-235 or plutonium-239 provide high-speed neutrons that can be used for bombardment reactions. A discussion of the structure of a nuclear reactor is given in a later section.

nuclear fission

Controlled bombardment of stable nuclei by neutrons in a reactor can produce radioactive isotopes. In these situations, the products are normally simple and predictable. With large unstable nuclei such as those of uranium-235, however, bombardment with neutrons gives a more complex mixture of products. These products result from the splitting of the immediate compound nucleus, uranium-236, into smaller nuclei whose mass numbers generally range from about 70 to 160. In addition to these fragments, several neutrons also are produced, which can bombard other uranium nuclei and thus perpetuate the reaction as a chain process. (See Figure 5-1, on page 86.) This process is known as **nuclear fission.**

(*On opposite page*)
Van de Graaf particle accelerator at Brookhaven National Laboratory, Upton, Long Island. Ions of hydrogen, chlorine, sulfur, and oxygen are among those that this device can accelerate to high speeds. (*Courtesy Brookhaven National Laboratory*)

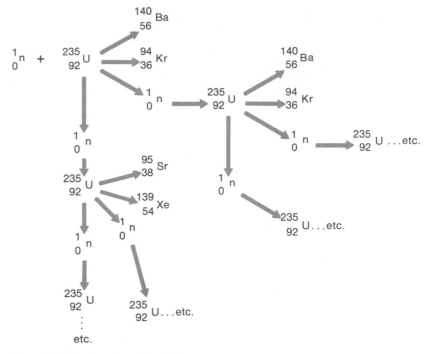

Figure 5-1. Fission of uranium-235

Fission processes were observed by Enrico Fermi in the late 1930s, and it was he who engineered the construction of the first atomic pile or reactor, which was built beneath the football stadium of the University of Chicago during World War II. In 1939, Otto Hahn and co-workers were the first to identify the products of such reactions. The more common isotope of uranium, uranium-238, does not undergo the fission process, but does convert to plutonium-239, which is fissionable.

$$^{238}_{92}\text{U} + {}^{1}_{0}n \rightarrow {}^{239}_{92}\text{U} \rightarrow {}^{239}_{93}\text{Pu} + {}^{0}_{-1}e$$

This process has become attractive as a potential source of fissionable material for nuclear reactors, since uranium-235, the fissionable isotope of uranium, is found only to the extent of 0.77% in natural uranium. A recently developed reactor, the fast breeder reactor, uses fast neutrons from a small initiator core of uranium-235 to convert uranium-238 into plutonium-239. The plutonium isotope then serves as the fissionable fuel in the reactor for energy generation.

The reactor is called a "fast breeder" reactor because it produces

The hazards of atmospheric testing of atomic weapons have been recognized for a number of years. As a consequence, the United States, Russia, and most of the world's smaller countries have signed an agreement not to conduct additional atmospheric testing. A few countries, however, have refused to sign the agreement. France and Red China, in particular, have continued tests of nuclear weapons. Both countries have been strongly criticized for their actions.

West stands of Stagg Field at the University of Chicago. Under these football stands the world's first self-sustained, controlled nuclear fission took place on December 2, 1942. (*Courtesy Argonne National Laboratory*)

The achievement of a controlled nuclear chain reaction was first realized in December, 1942, by Enrico Fermi and his team of physicists in Chicago. Fermi, a theoretical physicist from Italy, became interested in producing artificial radioactivity by neutron bombardment of stable isotopes. His work led him to the idea that perhaps a self-sustaining nuclear chain reaction might be developed to produce nuclear energy. He designed a device for this purpose, called an atomic pile, which was constructed on an abandoned squash court under the west stands of the University of Chicago's stadium. On December 2, 1942, six weeks after its construction was begun, the pile was operated for the first time and proved to be successful. The program that consolidated all the early experimental research in nuclear energy in the United States was called the Manhattan Project.

a greater quantity of fissionable material than is consumed in generating energy. The reason is that although the fission of one atom of plutonium-239 requires only one neutron per atom, the process produces several neutrons, some of which can be used for converting more uranium-238 to plutonium-239. Therefore, each plutonium-239 atom that undergoes fission can produce, on breeding, several additional atoms of plutonium-239 from uranium-238. The breeder process permits uranium-238, which has not been usable despite its

Argonne National Laboratory's Experimental Breeder Reactor II at the National Reactor Testing Station near Idaho Falls, Idaho. (*Courtesy Argonne National Laboratory*)

Although the mass loss in a nuclear reaction is very small in comparison with the total amount of matter involved, the amount of energy produced is extremely large. Einstein's equation shows that if a single gram of mass were completely converted to heat energy, 9×10^{20} ergs or 2.15×10^{10} kcal of heat would be produced. This amount of energy would be equivalent to that released by the combustion of about 35,500 tons of coal or 500,000 gallons of gasoline.

abundance, to serve as a source of fissionable material in a reactor. Another breeder process that is being developed involves the conversion of a relatively stable isotope of thorium, thorium-232, into uranium-233, which is fissionable. There is much interest in fast breeder reactors because of their great potential use in power plants for the production of electrical energy.

When a fission reaction occurs, tremendous amounts of energy are released. This energy is due to the conversion of a small amount of matter into energy. Albert Einstein expressed this conversion

mathematically with his well-known equation, $E = mc^2$, where E represents energy, m represents mass, and c represents the speed of light. As much as 0.1% of the starting material can be transformed into energy in a fission process.

nuclear reactors and power plants

The fission process is being used to generate electrical energy, principally through the device of the nuclear reactor. A reactor contains cylinders of materials, generally uranium-235 or plutonium-239, which are sources of neutrons. These cylinders are imbedded in graphite or heavy water (deuterium oxide), which serves as a moderator. The moderator slows the emitted neutrons so that they may better be able to bombard other atoms of uranium or plutonium. As a further control of the fission reaction, a reactor contains rods of cadmium or boron that have a high ability to absorb neutrons. These rods reduce or increase the number of neutrons that can be made

San Onofre Nuclear Generating Station near San Clemente, California. This nuclear power plant began commercial operation in January, 1968, and has a generating capacity of 450,000 kilowatts. (*Courtesy Southern California Edison Company*)

Left: Reactor vessel head is lowered into position over control rod drive shafts in a nuclear power plant. *Right*: Control room of a nuclear power plant. (*Courtesy left, Wisconsin Electric Power Company*; *right, Rochester Gas & Electric Company*)

available to continue the chain reaction in the reactor. Without the control rods, overproduction of neutrons results, which can lead to an uncontrolled fission reaction. The heat produced by the fission process can generate electrical power. This already has been done in nuclear-powered ships and some power plants.

environmental impact of nuclear power plants

There has been much concern lately about the depletion of the earth's natural fuels such as petroleum, natural gas, and coal. The increasing demand for energy, particularly electrical energy, derived from these fuels has led to the prediction that by the year 2000 most of these natural fuels will be nearly depleted. Alternative new sources of energy are being sought to relieve the large drain on natural fuels. One source that has been considered is nuclear fission. The construction and use of nuclear power plants raise several serious matters of concern, however.

First, how safe can nuclear power plants be made? Potential accidents could prompt release of deadly radioactive materials into the environment. Power-plant construction would need to incorporate several lines of defense and safety to assure minimum danger to the surrounding area. Proponents of nuclear power maintain that such safety and defense measures indeed can be taken. Furthermore, by constructing these plants in remote areas, away from populated centers, exposure of people to the small amount of radioactive material that routinely is released to the environment can be reduced.

Although nuclear power plants would not be expected to emit pollutants such as sulfur dioxide into the atmosphere, another kind of pollution must be recognized. This is **thermal pollution,** impairment of the environment by heat. The abnormal temperature conditions that this pollution could produce are harmful not only to fish but also to the entire aquatic ecological system. Operation of a nuclear plant generates large amounts of heat that must be removed by a cooling system. The most likely cooling system conducts heat away by flowing water diverted from some supply such as a river. The heated water from the cooling system is discharged back into the river, causing thermal pollution of the surroundings. Current technology is being applied to improve the structure of reactors and power plants, to diminish the amount of waste heat produced. Ways to utilize the waste heat are also under consideration.

A final major concern regarding the use of nuclear reactors to generate power is the disposal of the radioactive wastes and residues produced. Periodically, these wastes must be removed from the reactors, transported to a storage site, and placed in a repository strong enough to last for extremely long periods without allowing any leak of radioactive material. If the number of nuclear power plants is to be increased significantly, the problem of waste disposal becomes immense. Presently, no feasible solution exists other than maintaining the strictest controls and safety measures.

It appears that nuclear fission as an energy source poses major problems that must be solved before wide-scale use of nuclear power plants can become practical. One other type of nuclear process looks promising as an energy source, however. It has the advantage of being nonradioactive and, gram for gram, it would produce more energy than uranium fission. This process is **nuclear fusion,** the merging of small nuclei to form larger ones.

nuclear fusion

Nuclear fusion is believed to be the process by which the sun and stars produce their energy. It is now fairly well accepted that a series of nuclear reactions occurs in the sun that results in the combination and transformation of hydrogen into helium. This series of reactions can be summarized by the following equation.

$$4\,{}_{1}^{1}\text{H} \rightarrow {}_{2}^{4}\text{He} + \text{energy}$$

Some support for this series of nuclear conversions has been provided by detection of both hydrogen and helium in the sun.

A fusion reaction requires an extremely high temperature for initiation, but once started, the reaction can sustain itself by the tremendous amount of heat that it produces through conversion of some mass into energy. In the hydrogen bomb, which releases its power and energy through a fusion reaction, deuterium (^2_1H) is converted to helium. The high temperature of several million degrees Celsius needed to initiate the fusion process is provided by a small atomic bomb, which operates through nuclear fission.

Since the fusion process does not generate radioactive products, it has appeared attractive as an alternative source of nuclear energy. Deuterium is available in nature in approximately one out of every 6000 atoms of hydrogen. Appropriate separation techniques are known already and can yield large amounts of deuterium from the earth's oceans—enough to supply mankind with energy for the foreseeable future.

The powerful release of energy from a hydrogen bomb is the result of a nuclear-fusion reaction that generates helium from hydrogen. (*Courtesy U.S. Air Force*)

The amount of energy that can be obtained by fusion of even small amounts of hydrogen is enormous. If all the hydrogen atoms in a 250-gram sample of water (about a glassful) were to undergo a nuclear-fusion reaction, enough energy would be produced to convert 7000 tons of water, originally at room temperature (25°C), into steam.

Before this hydrogen isotope can be used as an energy source through fusion reactions, however, solutions must be found for technological problems associated with creating and controlling the high temperatures needed to initiate the reaction. It is also necessary to develop suitable techniques for containing the deuterium satisfactorily at this high temperature, to make it available for the fusion process. Some promising developments in hydrogen-fusion technology have been made in recent years, and it is possible that before the decade is over, fusion power plants may be producing electricity.

radioactive fallout

Much concern has been expressed about radioactive materials in the atmosphere. These materials were put into the environment principally by detonation of nuclear weapons in the atmosphere. Although major testing of atomic devices in the atmosphere now has been replaced largely by underground testing, enough radioactive material already has been released to contaminate the atmosphere for several more decades. Furthermore, widespread use of nuclear reactors as energy sources in power plants and atomic-powered ships could add to this contamination.

Several radioactive isotopes are found in the atmosphere, including strontium-90 and cesium-137. Through natural processes, these isotopes can be removed from the atmosphere as radioactive fallout and settle on the earth's surface. They can be assimilated into different life forms and, if sufficiently concentrated, can bring about radioactive damage to the organisms. Strontium-90, for example, can be taken up easily by cows through grass contaminated by radioactive fallout. Because of the chemical similarity of strontium to calcium, an important constituent in bones and milk, the strontium-90 atoms accumulate in the bones and milk of cows and then are transferred to humans who consume the milk. Nursing mothers also may transfer this isotope to their babies through breast feeding. Strontium-90 can cause bone deformations and bone cancer in animals and humans, owing to accumulation of the isotope in bone tissue.

Radiation burns on the back of a Nagasaki atomic-bomb victim correspond to the dark-colored areas of the kimono she was wearing at the time of the explosion. (*Courtesy U.S. Atomic Energy Commission*)

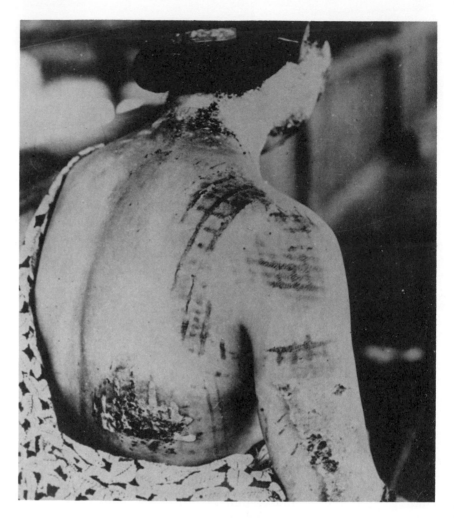

Through other natural food chains that exist among plants and animals, isotopes from radioactive fallout can accumulate and concentrate to abnormal levels in a number of life forms, including man. Damage to cell nuclei by radiation from these isotopes could lead to the formation of cancerous cells or to mutations through genetic alterations.

uses of nuclear reactions and radioactivity

Despite some of radiation's adverse effects, radioactive isotopes, nuclear reactions, and radiation produced by such processes have many beneficial functions in medicine, industry, and science. A few of these have been mentioned already, and some additional ones will be discussed in this section.

One of the outstanding applications of nuclear reactions is in the field of synthetic elements. Of the 105 elements currently known to scientists, only about 90 occur naturally. The remaining 15 or so are synthetic, produced in most cases by induced nuclear reactions. Some examples of these synthetic elements and the means by which they were produced are represented below.

$$^{238}_{92}U + {}^{1}_{0}n \rightarrow {}^{239}_{92}U \rightarrow {}^{239}_{93}Np + {}^{0}_{-1}e$$

$$^{239}_{93}Np \rightarrow {}^{239}_{94}Pu + {}^{0}_{-1}e$$

$$^{253}_{99}Es + {}^{4}_{2}He \rightarrow {}^{256}_{101}Md + {}^{1}_{0}n$$

$$^{246}_{96}Cm + {}^{12}_{6}C \rightarrow {}^{254}_{102}No + 4\,{}^{1}_{0}n$$

Testing and determination of the quality of materials is aided greatly by using radioactivity. Measurement of the density and thickness of materials can be accomplished with it, as can detection of flaws in metal products. Radioactive sterilization of foods permits

Elements present in ancient pottery are detected and identified by neutron activation analysis. (*Courtesy Steve Gerber, Lawrence Radiation Laboratory*)

their safe storage at room temperature for longer periods, although adverse effects on flavor, texture, and vitamin content have made this practice no longer widespread. Small quantities of elements in materials can be detected and measured by bombarding the materials with neutrons and analyzing the radiation emitted. The technique is called **neutron activation analysis.**

Radioactive isotopes are used widely in medicine and science as tracers in studying various phenomena. Because these isotopes emit detectable radiation, their location and travel in a living or chemical system can be determined easily. Some areas of study in which radioactive materials have been used are body functions and ailments, plant photosynthesis, cancer therapy, chemical reaction mechanisms, and drug metabolism.

nuclear versus electronic chemical reactions

Although the reactions that an atomic nucleus can undergo are significant and important, for most elements such changes generally are not considered to be the principal or standard reactions. So-called "ordinary" chemical reactions involve changes in the electronic arrangement in an atom, and these constitute the subject of the next three chapters.

study questions

1 How does a nuclear reaction differ from the more common type of chemical reaction that an atom undergoes?
2 What are the compositions of alpha and beta particles?
3 How would you determine the age of an archaeological specimen?
4 How would you determine the age of a rock?
5 How could the nuclear fission of uranium be employed to obtain usable energy?
6 Of what use are the various accelerators, such as the cyclotron and the betatron, in nuclear chemistry?
7 Compare the chemical processes occurring in an atomic bomb with those occurring in a hydrogen bomb.
8 Why is a fast breeder reactor so named? Why is it possible to generate more fuel than is consumed in a fast breeder reactor?

9 What problems would have to be solved if nuclear power plants were to become the major sources of energy?

10 Discuss the source of radioactive fallout.

11 List five peaceful uses of nuclear reactions.

special problems

1 The following isotopes are alpha emitters. Write an equation to illustrate this reaction for each isotope.
   a. $^{209}_{83}Bi$ b. $^{231}_{91}Pa$

2 The following isotopes are beta emitters. Write an equation to illustrate this reaction for each isotope.
   a. $^{32}_{15}P$ b. $^{90}_{38}Sr$

3 Complete the following equations.
   a. $^{197}_{80}Hg + {}^{0}_{-1}e \rightarrow$ _____
   b. $^{27}_{13}Al + {}^{4}_{2}He \rightarrow$ _____ $+ {}^{1}_{0}n$
   c. _____ $+ {}^{1}_{1}H \rightarrow {}^{6}_{3}Li + {}^{4}_{2}He$
   d. $^{45}_{21}Sc +$ _____ $\rightarrow {}^{45}_{20}Ca + {}^{1}_{1}H$
   e. $^{209}_{83}Bi +$ _____ $\rightarrow {}^{210}_{84}Po + {}^{1}_{0}n$

4 The half-life of $^{117}_{50}Sn$ is 14 days. How much of a 1-gram sample of this isotope would remain after 8 weeks?

chapter **6**

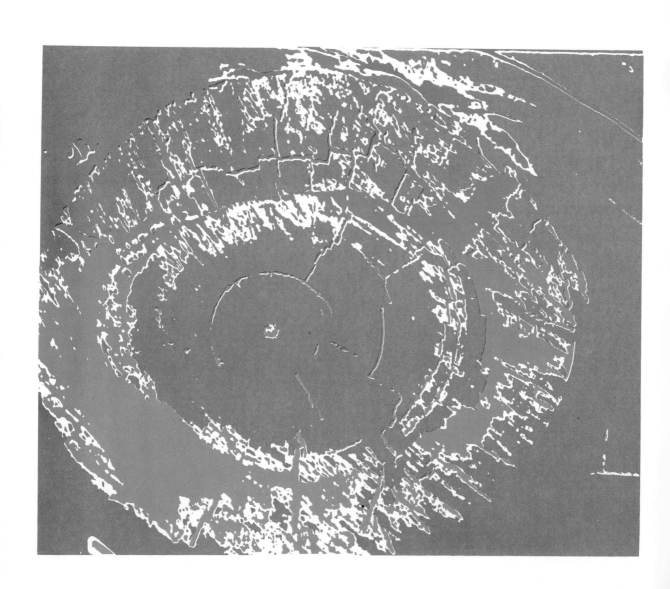

electronic structure of the atom

The Rutherford model provided the gross structural features of the atom. According to this model, electrons occupied the bulk of the atom's space. No information was provided concerning the arrangement of these electrons, however. Surprisingly, an understanding of the organizational pattern of electrons in the atom evolved from observations concerning the interaction of light with matter.

continuous spectrum
emission spectrum
absorption spectrum
ground state
excited state
quanta
principal energy level
orbital
s orbital
p orbital
d orbital
f orbital
electron configuration

emission spectra

For many years it has been known that compounds of certain metals will burn with brightly colored flames. The color is characteristic of the metal portion of the salt and has been used as a qualitative test for these metals. Table 6-1 lists a few metals and the corresponding color given in a Bunsen-burner flame.

Table 6-1. Some selected metals and their characteristic flame colors

Lithium	red	Cesium	blue
Sodium	yellow	Calcium	orange red
Potassium	violet	Strontium	brick red
Rubidium	red	Barium	green

An interesting novelty available in shops consists of wood chips that have been soaked in solutions of one or more of these compounds. When the chips are placed in a fireplace, the flames become tinted with the characteristic colors of the various metals.

A more precise analysis of the light emitted by these elements was provided in 1859, with the development of the spectroscope. A simple spectroscope is outlined in Figure 6-1. In this instrument, the

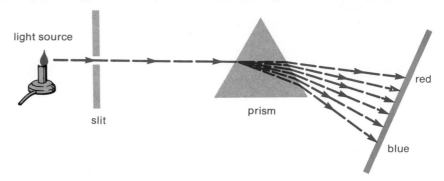

Figure 6-1. Light path in a simple spectroscope

light from a source is restricted to a narrow beam by a slit. This narrow beam then passes through a prism, which is an optical device that bends light passing through it. The extent of bending relates to the wavelength of light. Shorter wavelengths are bent more than longer wavelengths. White light emerges as a **continuous spectrum** like the colors in a rainbow. (Please refer to the full-color illustrations inside the front cover.)

When a chemical element is heated and used as a light source for a spectrascope, a series of lines is seen. This constitutes an **emission spectrum**, because it represents the light emitted by the element. It is so characteristic that the elements rubidium, cesium, thallium, indium, gallium, and scandium were discovered as a result of spectroscopic examination of their minerals. These minerals showed unexplained lines in their spectra that later turned out to be due to the elements listed.

The elemental compositions of the sun and stars also have been determined by this technique. Helium was discovered during a study of the spectrum of the sun in 1868; the element was not identified on earth until 1895.

absorption spectra

When solids are heated to very high temperatures, they emit light of all visible wavelengths. Light of this type is said to have a *continuous* emission spectrum. (Again please refer to the full-color illustrations inside the front cover.) When this light passes through certain substances, selected wavelengths of the light are absorbed. If the transmitted light then is passed through a prism, as in a spectroscope, the absorbed wavelengths are missing and appear as dark

Presently in the United States, helium is obtained as a minor constituent of natural gas. Much of this natural resource is wasted, however, as the demand for natural gas far exceeds the demand for helium. For a time, the federal government supported a program to separate and store much of this helium for later use. The government has since abandoned the program as part of its cost-cutting activities. When natural gas supplies are exhausted, helium will have to be obtained from the atmosphere, just as similar gases are. The resulting price increase will affect both scientific studies, such as those using gas chromatography, and everyday activities, such as those dependent on products made by heliarc welding.

lines in the spectrum. The dark lines in the **absorption spectrum** are found at the same wavelength as the light lines in the corresponding emission spectrum.

Many elements and compounds also absorb light of wavelengths that are too long (infrared) or too short (ultraviolet) to be seen. These characteristic absorption patterns are used in analytical and theoretical studies. The entire scale of wavelengths is known as the electromagnetic spectrum. Figure 6-2 shows the complete electromagnetic spectrum with the various regions labeled.

Figure 6-2. The electromagnetic spectrum, showing the ranges of wavelengths comprising the different types of radiation.

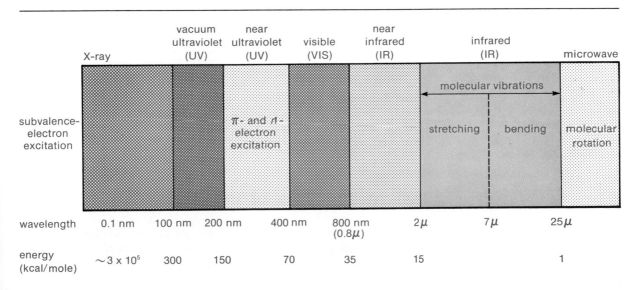

Figure 6-3. Fluorine, according to the Bohr model

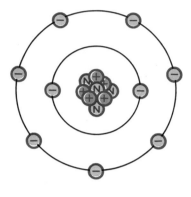

Figure 6-3. Fluorine, according to the Bohr model

the bohr atom

The spectral properties of elements created a major dilemma for atomic theory. Existing models of the atom could not explain this property. It was not until 1913, when Niels Bohr suggested a new model, that this behavior became understandable.

Bohr visualized the atom as consisting of a dense, small nucleus surrounded by electrons in definite orbits. Each orbit varied in energy, depending on its proximity to the nucleus, with the closer orbits having lower energy. This scheme is illustrated in Figure 6-3. This model is sometimes called the "solar system model" because of the similarity, in relative locations, of the sun to the nucleus and the planets to the electrons. Each orbit is given a letter designation, K, L, M, N, and so on, beginning with the lowest-energy orbit.

Bohr proposed that, under normal conditions, electrons exist in orbits of relatively low energy. He called these **ground states**. When an atom is exposed to a source of high energy, such as heat or light, its electrons are forced into orbits of higher energy, which Bohr called **excited states**. Bohr proposed that, when electrons in the excited state fall back to the ground state, they release their excess energy in the form of light. The wavelength of this emitted light depends on the amount of energy involved. Since the wavelength (or frequency) of absorption or emission is constant for a given element, Bohr proposed that the electron is excited to and returned to definite energy levels. This proposal was consistent with the work of Max Planck who, in 1900, had proposed that light consists of units of energy called **quanta**. The energy of a quantum of light is proportional to the frequency of the light.

According to the Bohr theory, when an atom absorbs energy an electron jumps to a higher-energy orbit. Since there are only a few discrete orbits that an electron can occupy, there are only a few discrete amounts of energy that can be absorbed, leading to a discrete absorption band. Similarly, when an excited atom loses energy, the electrons fall back to the ground state orbits. Again, only a few discrete energy levels are possible, leading to discrete emission bands. This process is represented in Figure 6-4.

Figure 6-4. Absorption and emission of energy by a Bohr atom

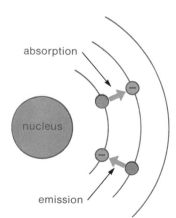

absorption

nucleus

emission

modern concepts of the atom

Much interest was generated in the area of atomic spectra as a result of the Bohr concept. As a consequence, four different subenergy levels soon were discovered. These subenergy levels, represented by bands in the spectra, were labeled *sharp, principal, diffuse,* and *fundamental,* and were given the abbreviations *s, p, d,* and *f.* Sub-energy bands are inconsistent with the Bohr atom structure, however. Prior to this discovery, the Bohr atom structure appeared to accommodate the known data remarkably well—so well, in fact, that it quickly was adopted almost universally and was displaced only slowly by the quantum-mechanical concept of the atom.

The atomic theory commonly accepted today results from the complex quantum-mechanical calculations of Schroedinger and other researchers. Much of this work is based on the Schroedinger equation, which is a complex mathematical representation of the atom developed in 1926. The solution to this equation predicts that there should be a series of **principal energy levels** corresponding to those of the Bohr atom. In addition, a series of subenergy levels within each major level is predicted, corresponding to the *s, p, d,* and *f* bands observed in spectra.

The mathematics of these calculations are much too complicated even for many scientists and will not be considered here. It is possible, however, to gain a reasonable understanding of the theory from a qualitative discussion of the results of such calculations.

principal energy levels

As with the Bohr atom, the quantum-mechanical model of the atom identifies a series of principal energy levels that contain the electrons. In the Bohr atom, these shells or orbits were labeled *K, L, M, N,* and so forth. In the quantum-mechanical model, they are known simply as principal energy levels and are labeled 1, 2, 3, 4, 5, and so on. Thus, for example, the *K* shell corresponds to the first principal energy level, and the *L* shell corresponds to the second principal energy level. Table 6-2 shows the correspondence between the Bohr

Table 6-2. Principal energy levels and electrons

	K	*L*	*M*	*N*	*O*
Bohr Atom Shell					
Principal Energy Level	1	2	3	4	5
Maximum Number of Electrons	2	8	18	32	50

atom shells and the quantum-mechanical principal energy levels. The table also gives the maximum number of electrons that may occur in each energy level.

The electrons in each principal energy level of the quantum-mechanical atom are located in various subenergy levels discussed in the next section. These subenergy levels correspond to the spectroscopically observed s, p, d, and f levels.

atomic orbitals

A principal feature of the quantum-mechanical model of the atom is the introduction of the concept of subenergy levels, or **orbitals**. An orbital is a region in space around the nucleus that has a high probability of containing an electron at any given moment. In other words, orbitals do not define the path of an electron but merely define its probable location in space. These orbitals correspond to the subenergy levels described earlier and thus are labeled s, p, d, and f. According to modern theory, each orbital may contain a maximum of two electrons.

The lowest-energy orbital of any principal energy level is called an **s orbital**, which has a spherical shape with the nucleus at the center of the sphere. Figure 6-5 gives a drawing of an s orbital. There is one s orbital for each principal energy level in the atom. The s orbital of the first principal energy level is called the $1s$ orbital. The s orbitals of the successive principal energy levels are designated $2s$, $3s$, $4s$, and so on.

The next higher-energy orbital within a given principal energy level is called a **p orbital**. It has a dumbbell shape with the nucleus between the two lobes. There are three different p orbitals per major energy level, but each has the same energy. The difference among the three p orbitals lies not in their shape or energy but rather in their orientations in space. These relative orientations are illustrated in Figure 6-6. One orbital lies in the horizontal direction in the plane of the paper and is called the p_x orbital. The p_y orbital lies in the vertical direction in the plane of the paper and is perpendicular to the p_x orbital. The p_z orbital is perpendicular to both the paper and the other two p orbitals. It should be emphasized that these orbitals are all oriented around a single, common nucleus.

Since the first principal energy level can hold only two electrons, it does not have any p orbitals. The second principal energy level can hold a maximum of eight electrons and therefore contains p orbitals in addition to an s orbital. Thus the lowest-energy p orbitals in an

Figure 6-5 An s orbital

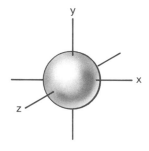

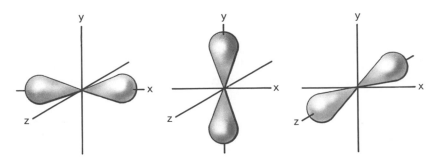

Figure 6-6. Various *p* orbitals of a principal energy level

atom are 2*p* orbitals. The next higher-energy orbitals of the *p* variety are the 3*p*, 4*p*, 5*p*, and 6*p* orbitals.

There are five different *d* orbitals and seven different *f* orbitals that may be found in any one principal energy level. The *d* orbitals are not found until the third principal energy level in an atom, so the lowest-energy orbital of this type is the 3*d* orbital. The lowest level in which *f* orbitals are found is the fourth principal energy level, and these are labeled 4*f* orbitals. The shapes of these *d* and *f* orbitals are more complex than those of the *s* and *p* orbitals. In principle, there also should be orbitals above the *f* orbitals. Actually, however, these orbitals are unknown, because all known natural and synthetic elements contain too few electrons to require these orbitals.

The relative energies of the various orbitals are shown in Figure 6-7. The energy at which each principal energy level begins has the

Figure 6-7. Relative energies of orbitals

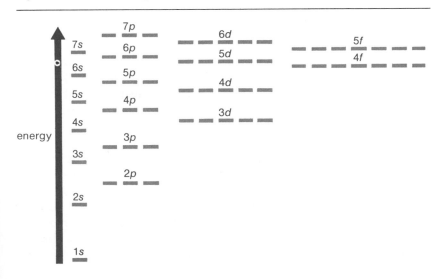

greatest difference between the first and second principal levels. With the third level, the principal energy levels begin to overlap. Thus the 4s orbital is lower in energy than the 3d orbital. As the energy difference between the principal energy levels becomes smaller, more overlapping is observed.

A memory device for predicting the order of orbitals

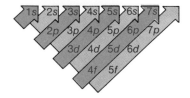

A simple memory device for predicting the order of orbitals in terms of increasing energy is shown in the accompanying illustration. Each of the orbital types is written in increasing order. Diagonal lines are then drawn through the orbitals as indicated. Thus the order is predicted correctly to be 1s 2s 2p 3s 3p 4s 3d 4p 5s, and so forth.

filling orbitals

The distribution of electrons in orbitals within an atom is called the atom's **electron configuration**. One can now describe electron configuration beginning with hydrogen, with atomic number 1, and proceeding by increasing atomic number until all elements have been considered. Normally the electrons will go into an available orbital of lowest energy first, in the order shown in Figure 6-7. In other words, the 4s orbital will be filled before the 3d orbital. Several examples should illustrate the procedure adequately.

Hydrogen is the smallest atom, with an atomic number of 1. Therefore, a hydrogen atom contains one electron and one proton. The lowest-energy orbital in an atom is the 1s orbital. Thus the single hydrogen electron is found in the 1s orbital and is labeled a 1s electron. Normally the electron configuration for hydrogen is designated simply as $1s^1$. The superscript indicates that the 1s orbital contains one electron.

Helium is the next larger atom, with an atomic number of 2. Therefore a helium atom contains two electrons. Since the 1s orbital possesses the lowest energy and can hold two electrons, both helium electrons are found in the 1s orbital. The electron configuration for helium normally is designated $1s^2$.

The lithium atom contains three electrons. The first two electrons are placed in the 1s orbital, as with helium. Since this orbital can hold only two electrons, the third electron now must be placed in the next higher-energy orbital. From Figure 6-7, it is seen that this would be the 2s orbital. The electron configuration of lithium then is designated as $1s^2 2s^1$. Since the inner portion of the electron arrangement in a lithium atom is identical to that of the helium atom, the electron configuration of lithium sometimes is abbreviated $[He]2s^1$.

The second principal energy level contains two kinds of orbitals, s and p. The p orbitals begin to fill with boron, whose electron configuration is $1s^2 2s^2 2p^1$ or $[He]2s^2 2p^1$.

Carbon, with six electrons, introduces a further consideration. The electron configuration of this element will be $1s^2 2s^2 2p^2$ or $[He]2s^2 2p^2$. The question is whether both p electrons occupy the same p orbital or whether they go into separate p orbitals, thus occupying two of the three available p orbitals. Following the rule that the electrons will try to assume the lowest-energy configuration, each electron will go into a separate p orbital. The reason is that electrons tend to repel each other, owing to their like charges. Thus, in the lowest-energy configuration, electrons would occupy different regions of space or, in this situation, separate p orbitals. This information is not represented, however, in configurations written for the various elements.

The electron configurations of the first 30 elements are listed in Table 6-3. It may be observed that elements containing d orbitals do not always rigidly follow the rules of electron distribution. Chromium and copper are two examples. These exceptions are complex and will not be discussed.

The electron configurations of atoms provide a way of organizing the different elements into groups or families of elements. This arrangement leads to the periodic table, which is discussed further in the next chapter.

Table 6-3. Electron configurations of the first 30 elements

Element	Atomic Number	Electron Configuration
H	1	$1s^1$
He	2	$1s^2$
Li	3	$1s^2 2s^1$
Be	4	$1s^2 2s^2$
B	5	$1s^2 2s^2 2p^1$
C	6	$1s^2 2s^2 2p^2$
N	7	$1s^2 2s^2 2p^3$
O	8	$1s^2 2s^2 2p^4$
F	9	$1s^2 2s^2 2p^5$
Ne	10	$1s^2 2s^2 2p^6$
Na	11	$1s^2 2s^2 2p^6 3s^1$
Mg	12	$1s^2 2s^2 2p^6 3s^2$
Al	13	$1s^2 2s^2 2p^6 3s^2 3p^1$
Si	14	$1s^2 2s^2 2p^6 3s^2 3p^2$
P	15	$1s^2 2s^2 2p^6 3s^2 3p^3$
S	16	$1s^2 2s^2 2p^6 3s^2 3p^4$

Table 6-3. Electron configurations of the first 30 elements (*continued*)

Element	Atomic Number	Electron Configuration
Cl	17	$1s^2 2s^2 2p^6 3s^2 3p^5$
Ar	18	$1s^2 2s^2 2p^6 3s^2 3p^6$
K	19	$1s^2 2s^2 2p^6 3s^2 3p^6 4s^1$
Ca	20	$1s^2 2s^2 2p^6 3s^2 3p^6 4s^2$
Sc	21	$1s^2 2s^2 2p^6 3s^2 3p^6 4s^2 3d^1$
Ti	22	$1s^2 2s^2 2p^6 3s^2 3p^6 4s^2 3d^2$
V	23	$1s^2 2s^2 2p^6 3s^2 3p^6 4s^2 3d^3$
Cr	24	$1s^2 2s^2 2p^6 3s^2 3p^6 4s^1 3d^5$
Mn	25	$1s^2 2s^2 2p^6 3s^2 3p^6 4s^2 3d^5$
Fe	26	$1s^2 2s^2 2p^6 3s^2 3p^6 4s^2 3d^6$
Co	27	$1s^2 2s^2 2p^6 3s^2 3p^6 4s^2 3d^7$
Ni	28	$1s^2 2s^2 2p^6 3s^2 3p^6 4s^2 3d^8$
Cu	29	$1s^2 2s^2 2p^6 3s^2 3p^6 4s^1 3d^{10}$
Zn	30	$1s^2 2s^2 2p^6 3s^2 3p^6 4s^2 3d^{10}$

study questions

1 Compare the appearance of a continuous spectrum with that of an emission spectrum.
2 Why does white light produce a continuous spectrum when passed through a prism?
3 How may absorption or emission spectra be used to detect the presence of elements?
4 What are the inadequacies of the Bohr concept of the atom?
5 What are the maximum numbers of electrons that may be contained in each of the first four principal energy levels?
6 What is the maximum number of electrons that can occupy an *s* orbital? A *p* orbital? A *d* orbital?
7 What are the maximum numbers of *s*, *p*, *d*, and *f* electrons that may be found in any one principal energy level?
8 What are the maximum numbers of *s*, *p*, *d*, and *f* orbitals that may be found in a principal energy level?
9 Compare the shape of an *s* orbital with that of a *p* orbital.
10 Describe the orientation in space of the three *p* orbitals (p_x, p_y, and p_z) found in a principal energy level.
11 How does the energy of a $2p_x$ electron compare with that of a $2p_y$ electron?
12 Draw a picture of the carbon atom showing all its electron orbitals.

13 Indicate which orbital in each of the following pairs is higher in energy.
  a. 2s or 3s   d. 4f or 5d
  b. 5p or 5d   e. 6p or 7s
  c. 3d or 4s   f. 4d or 6s

14 Write out the complete electron configuration for each of the following elements.
  a. Li   c. N   e. Al
  b. B   d. Na   f. Cl

special problems

1 When glass tubing is heated in a flame, a yellow color is produced. To what element might this be due, and how is the color produced?

2 Write out the complete electron configuration for each of the following elements.
  a. K   c. As   e. Ba
  b. Fe   d. Cd   f. Pb

3 a. Write out the complete electron configurations for neon (Ne), argon (Ar), and krypton (Kr).
  b. These three elements are members of a family of elements called the noble gases. What similarity in the electron configurations of these elements do you observe?
  c. All noble gases have the chemical similarity that they do not readily form compounds. Suggest an explanation for this behavior based on a study of their electron configurations.

chapter **7**

the periodic table

Over the years, it became obvious that many elements were similar in their chemical properties. For example, sodium and potassium are both metals that react violently with water. Furthermore, they both form compounds with chlorine in a 1 to 1 atomic ratio ($NaCl$ and KCl). The periodic table is an arrangement of the elements in increasing atomic number, which places elements of similar chemical properties into families. Knowledge of the periodic table is useful in the study of chemistry because physical and chemical properties of an element are implied by the location of that element in the table.

periodic table
metal
nonmetal
noble (inert) gas

development of the periodic table

In the early 1800s, a number of scientists attempted to correlate the chemical and physical properties of the known elements. These scientists were hampered in their efforts because only a small number of elements were known then and the concept of atomic number had not evolved yet. Mendeleev, a Russian scientist, generally is credited with developing the first truly successful correlation. He arranged the elements in a table of increasing atomic weight, with elements of similar properties in vertical columns. By allowing chemical properties to take precedence over atomic weight whenever a conflict appeared, Mendeleev evolved a periodic table of the elements that is similar to that used today.

Since Mendeleev's time, several different periodic arrangements of the elements have been devised, to emphasize various important characteristics of the elements. This chapter gives several examples, including some unusual versions for comparison to the standard and for historical interest.

The most common way to construct the periodic table used today is based on the rules of electron configuration discussed in the

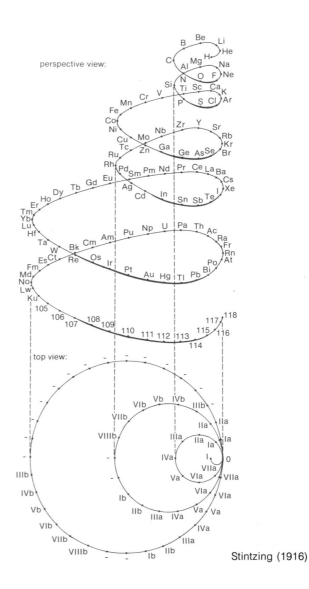

Li										Be	B	C	N	O	F
Na										Mg	Al	Si	P	S	Cl
K	Ca	-	Ti	V	Cr	Mn	Fe	Ni, Co	Cu	Zn	-	-	As	Se	Br
Rb	Sr	-	Zr	Nb	Mo	Rh	Ru	Pd	Ag	Cd	U	Sn	Sb	Te	I
Cs	Ba	-	-	Ta	W	Pt	Ir	Os	Hg	-	Au	-	Bi	-	-
Tl	Pb	-													

Mendeleev (1869)

perspective view:

top view:

Stintzing (1916)

Reproduced from *Graphic Representations of the Periodic System during One Hundred Years*, by Edward G. Mazurs. Copyright © 1974 by The University of Alabama Press. Used by permission of author and publisher.

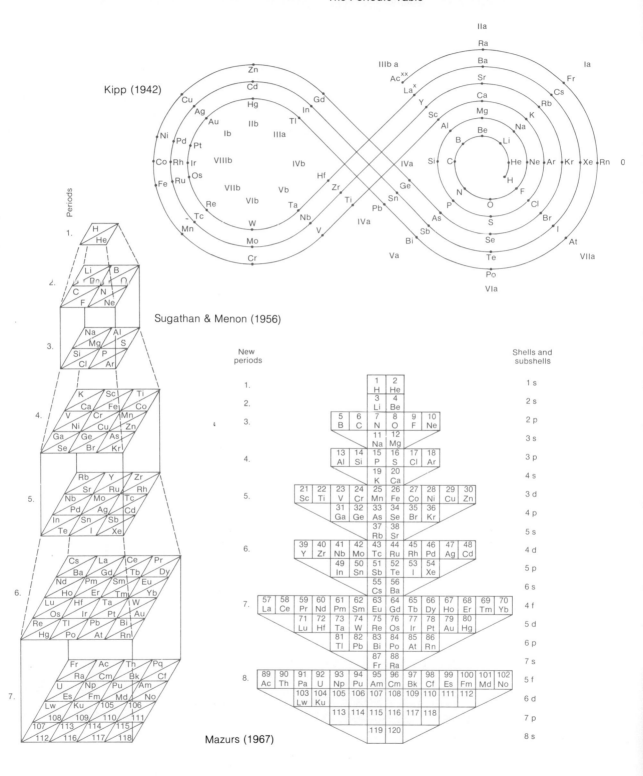

Kipp (1942)

Sugathan & Menon (1956)

Mazurs (1967)

previous chapter. A table constructed this way has elements of similar electron configuration in vertical columns and is like the table developed by Mendeleev. The vertical columns of Mendeleev's table contain the same elements as the corresponding columns in the electron configuration table. This similarity implies that electron configuration is of prime importance in determining the physical and chemical properties of an element. This is indeed true; electron configuration does dictate chemical and physical properties, as will be shown in the next chapter.

Figure 7-1 is a modern periodic table organized according to electron configuration. A periodic table usually contains the chemical symbol, the atomic number, and the atomic weight of each element. Additional information concerning the electronic structure, and consequently the chemical properties, of the elements may be deduced from the positions of the elements in the table. Other data such as crystal structures may sometimes be added to certain periodic tables.

Figure 7-1. Periodic table of the elements

IA	IIA	IIIB	IVB	VB	VIB	VIIB	VIII	VIII	VIII	IB	IIB	IIIA	IVA	VA	VIA	VIIA	noble gases
1 H 1.01																	2 He 4.00
3 Li 6.94	4 Be 9.01											5 B 10.8	6 C 12.0	7 N 14.0	8 O 16.0	9 F 19.0	10 Ne 20.2
11 Na 23.0	12 Mg 24.3											13 Al 27.0	14 Si 28.1	15 P 31.0	16 S 32.1	17 Cl 35.5	18 Ar 39.9
19 K 39.1	20 Ca 40.1	21 Sc 45.0	22 Ti 47.9	23 V 50.9	24 Cr 52.0	25 Mn 54.9	26 Fe 55.8	27 Co 58.9	28 Ni 58.7	29 Cu 63.5	30 Zn 65.4	31 Ga 69.7	32 Ge 72.6	33 As 74.9	34 Se 79.0	35 Br 79.9	36 Kr 83.8
37 Rb 85.5	38 Sr 87.6	39 Y 88.9	40 Zr 91.2	41 Nb 92.9	42 Mo 95.9	43 Tc 98.9	44 Ru 101	45 Rh 103	46 Pd 106	47 Ag 108	48 Cd 112	49 In 115	50 Sn 119	51 Sb 122	52 Te 128	53 I 127	54 Xe 131
55 Cs 133	56 Ba 137	57* La 139	72 Hf 178	73 Ta 181	74 W 184	75 Re 186	76 Os 190	77 Ir 192	78 Pt 195	79 Au 197	80 Hg 201	81 Tl 204	82 Pb 207	83 Bi 209	84 Po (210)	85 At (210)	86 Rn (222)
87 Fr (223)	88 Ra (226)	89** Ac (227)	104 Ku (257)	105 Ha (260)													

*	58 Ce 140	59 Pr 141	60 Nd 144	61 Pm (147)	62 Sm 150	63 Eu 152	64 Gd 157	65 Tb 159	66 Dy 163	67 Ho 165	68 Er 167	69 Tm 169	70 Yb 173	71 Lu 175
**	90 Th 232	91 Pa (231)	92 U 238	93 Np (237)	94 Pu (242)	95 Am (243)	96 Cm (247)	97 Bk (247)	98 Cf (249)	99 Es (254)	100 Fm (253)	101 Md (256)	102 No (254)	103 Lr (257)

(Numbers in parentheses represent mass numbers of the most stable isotopes)

general shape of the periodic table

The modern periodic table can be divided into four major areas. All elements in any one of these areas have this similarity: the highest-occupied atomic orbital in each element is of the same type. Thus, all elements in the area designated s have s orbitals as their occupied atomic orbital of highest energy. For example, the configuration of sodium is $[Ne]3s^1$ and that of calcium is $[Ar]4s^1$. The highest-occupied orbital in each of these elements is an s orbital. Likewise, elements in the d area all have d orbitals as their highest-occupied atomic orbitals. Zinc, for example, has the configuration $[Ar]4s^23d^{10}$. Similarly, elements in the p and f areas have p and f orbitals as their highest-occupied atomic orbitals. This feature of the periodic table is shown in Figure 7-2.

Figure 7-2. The shape of the periodic table as determined by orbital type

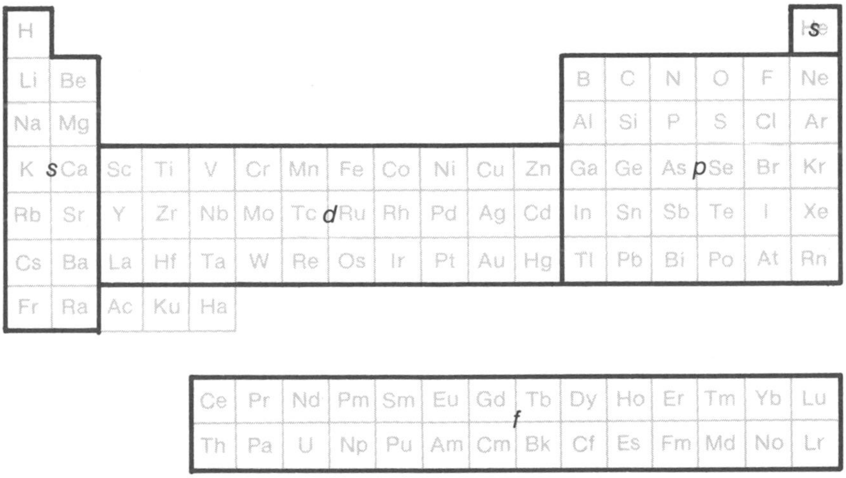

According to Figure 7-2, helium appears to be out of place. It is above the other noble gases on the right side of the table even though it only has $1s$ electrons. It would seem possible to place helium above beryllium in the table, owing to the similarity of the beryllium configuration ($[He]2s^2$) to that of helium. Helium, however, does not behave as the other elements do that contain two electrons in an outer s orbital, like beryllium or magnesium. Helium is placed with the noble gases instead because it behaves chemically as these elements do.

horizontal rows and principal energy levels

The principal energy level of the highest-occupied atomic orbital also can be determined through the periodic table, by looking at the horizontal row to which an element belongs, as indicated in Figure 7-3. For example, in the s region, the first row corresponds to the $1s$ orbital, the second row to the $2s$ orbital, and so forth. Since there are no $1p$ electrons, the first row in the p portion of the periodic table corresponds to the second principal energy level. Similarly, the d portion of the table begins with the third principal energy level and the f region begins with the fourth principal energy level.

Figure 7-3. Division of the orbital regions of the periodic table according to principal energy levels

1s																	1s
Li 2s Be												B	C	N 2p O	F	Ne
Na 3s Mg												Al	Si	P 3p S	Cl	Ar
K 4s Ca	Sc	Ti	V	Cr	Mn 3d Fe	Co	Ni	Cu	Zn	Ga	Ge	As 4p Se	Br	Kr
Rb 5s Sr	Y	Zr	Nb	Mo	Tc 4d Ru	Rh	Pd	Ag	Cd	In	Sn	Sb 5p Te	I	Xe
Cs 6s Ba	La	Hf	Ta	W	Re 5d Os	Ir	Pt	Au	Hg	Tl	Pb	Bi 6p Po	At	Rn
Fr 7s Ra	Ac	Ku	Ha											

Ce	Pr	Nd	Pm	Sm	Eu	Gd 4f Tb	Dy	Ho	Er	Tm	Yb	Lu
Th	Pa	U	Np	Pu	Am	Cm 5f Bk	Cf	Es	Fm	Md	No	Lr

The $3d$ level is between the $4s$ and $4p$ areas in the same horizontal row. This reflects the order of filling or the relative energies of the levels. In Chapter 6 it was shown that the $4s$ orbitals are relatively lower in energy than the $3d$ orbitals. For this reason the $4s$ levels are filled before the $3d$ orbitals. To maintain the steady progression in atomic number throughout the table, it is necessary to place the $3d$ row between the $4s$ and $4p$ rows.

vertical columns and chemical families

All elements in a given vertical column have the same type of orbital as their highest-occupied atomic orbitals. The only difference is in the principal energy level in which this orbital is found. Sodium, for example, has a $3s$ orbital as its highest-occupied orbital, and potassium has a $4s$ orbital as its highest-occupied orbital. Furthermore, each element in a given column also contains the same number of electrons in that outer orbital. Thus sodium and potassium both contain just one electron in their outer s orbitals.

If chemical behavior is related to electron configuration, as implied earlier, one might expect the chemical properties of sodium and potassium to be similar. This is indeed the case. Both metals are highly reactive and form strongly basic (alkaline) products with water.

This method of grouping the elements not only is convenient on the basis of atomic structure theory, but also leads to classification

THE DiGREGORIO PERIODIC CHART of CONTEMPORARY ELEMENTS

1 H HALDEMAN	2 He HEROIN																
3 Li LIDDY	4 Be BEEF											5 B BUGGING	6 C CYPRESS	7 N NAMATH	8 O OBSCENITY	9 F FOOTBALL	10 Ne NEGOTIATE
11 Na NASA	12 Mg MAGRUDER											13 Al ALASKA PIPELINE	14 Si STONEWALL IT	15 P PEACE	16 S SIRICA	17 Cl COST OF LIVING	18 Ar AARON
19 K KLEINDIENST	20 Ca CARSON	21 Sc SAN CLEMENTE	22 Ti THAILAND	23 V VASECTOMY	24 Cr CREDIBILITY	25 Mn McGOVERN	26 Fe FEMALE	27 Co COSELL	28 Ni NORTHERN IRELAND	29 Cu COVER UP	30 Zn ZEN	31 Ga GASOLINE	32 Ge GERMS	33 As AIR STAGNATION	34 Se SAM ERVIN	35 Br BIG BROTHER	36 Kr KISSINGER
37 Rb ROBBERY	38 Sr SHORTAGE	39 Y YOUTH	40 Zr ZIEGLER	41 Nb NEIGHBOR	42 Mo MONTREAL OLYMPICS	43 Tc TOBACCO	44 Ru RUSSIA	45 Rh REHABILITATE	46 Pd POLICE DEPARTMENT	47 Ag AGNEW	48 Cd CREDIT	49 In INFLATION	50 Sn SUBPOENA	51 Sb SCHOOL BOARD	52 Te TECHNOLOGY	53 I IMPEACH	54 Xe XEROX
55 Cs CRISIS	56 Ba BRALESS	57 La LAUGHTER	72 Hf HEFNER	73 Ta TAPES	74 W WATERGATE	75 Re REPUBLICAN	76 Os OIL SHORTAGE	77 Ir INTEREST RATES	78 Pt PHASE TEN	79 Au AUTOMOBILE	80 Hg HIJACKING	81 Tl TOPLESS	82 Pb PLAYBOY	83 Bi BICENTENNIAL	84 Po POLLUTION	85 At ARAB TERRORIST	86 Rn RICHARD NIXON
87 Fr FREEZE	88 Ra RACISM	89 Ac ARCHIBALD COX	104 Td TRICKY DICK	105 Ka KICK AROUND	106 A ANYMORE												

58 Ce CENSORSHIP	59 Pr PRICES	60 Nd NADER	61 Pm PERMISSIVE	62 Sm STOCK MARKET	63 Eu EUROPE	64 Gd GOLDWATER	65 Tb TIRED BLOOD	66 Dy DANDY DON	67 Ho HOMOSEXUAL	68 Er EHRLICHMAN	69 Tm TIME MAGAZINE	70 Yb YOUNG BLOOD	71 Lu LUTHER KING
90 Th THE END	91 Pa PAPACY	92 U UN-EMPLOYMENT	93 Np NOBEL PRIZE	94 Pu PRURIENT	95 Am AMNESTY	96 Cm CAMBODIA	97 Bk BLACK	98 Cf COFFEE	99 Es ESCALATION	100 Fm FEMININE	101 Md MIAMI DOLPHINS	102 No NOISE	103 Lr LABOR RELATIONS

When these elements are placed in their proper alignment, the following are noted:

DIAGONAL RULE: Elements 3 and 12; elements 8 and 17; elements 40 and 73; elements 53 and 86.

"SKIP ONE" RULE: Elements 3 and 5; elements 7 and 9; elements 10 and 36.

CONSECUTIVE RULE: Elements 73, 74, 75; elements 97 and 98; elements 81 and 82.

ANOMALIES: Not related: elements 29 and 86; elements 31 and 38; elements 49 and 77.

CURIOUS POINT: All the elements that allude to Watergate are either even-numbered or odd-numbered.

NOTE: Elements 104, 105, and 106, which have been prepared but are still not officially named by any group or individual of note, have been named by J. S. DiGregorio as of this date.

of the elements into chemical families. A few more general features of the periodic table can be pointed out at this time. A more detailed discussion will evolve as the concept of chemical bonding is developed in the next chapter.

The **metals** are located on the left side of the table, and the **nonmetals** are located on the right side. Metals are elements that typically have shiny luster, conduct electricity, and are malleable and ductile. Nonmetals normally lack luster, tend to be insulators rather than conductors of electricity, and are not malleable or ductile. There is a gradual transition from metallic to nonmetallic behavior across the periodic table. Elements in the middle of the table, where the *d* orbitals are being filled, are known as the *transition metals*. These elements are primarily metallic in nature but tend to have some intermediate properties. The dark line in the periodic table of Figure 7-1, which goes from the left side of boron (B) stepwise down to the left side of astatine (At), defines the generally accepted division between metallic and nonmetallic elements. Elements to the right of this line show little metallic character.

Elements on the far right of the table are known as the **inert** or **noble gases**. Except under special circumstances they are completely inert, failing to react with any other elements. The nature and significance of the electron configuration of the noble gases will be discussed later.

study questions

1 Who was Mendeleev, and what was his contribution to the study of elements?

2 What is the basis for the organization of the modern periodic table?

3 Would the same arrangement of elements be obtained if they were listed in a periodic table in order of increasing atomic weights rather than increasing atomic numbers? Why?

4 What uses can be made of the periodic table?

5 Using Figures 7-1 and 7-2, determine the orbital type (*s*, *p*, *d*, or *f*) for the highest-occupied atomic orbital in each of the following elements.

   a. H   d. Se   g. Eu   j. U
   b. Li   e. Xe   h. Sn
   c. B   f. Mg   i. Cd

6 Using Figures 7-1 and 7-3, determine the principal energy level for the highest-occupied atomic orbital for each of the elements in Question 5.

7 How many electrons are there in the highest-occupied atomic orbital of each of the elements in Question 5?

8 For each of the elements in Question 5, write the electron configuration for electrons beyond those corresponding to the preceding noble gas.

9 Would you expect the elements fluorine, chlorine, bromine, and iodine to have similar chemical properties? Why?

10 Would the elements nitrogen, oxygen, and fluorine have similar chemical properties? Why?

11 Give five examples of metals and five examples of nonmetals.

12 Generally, where in the periodic table are the nonmetals found?

chapter **8**

chemical bonding
and compound formation

Since chemical compounds do not resemble the elements from which they are made, they must be considered new, unique substances. Whenever two or more elements interact to form a compound, the electron arrangements in the outer energy levels are disturbed. This disturbance results in chemical linkages between atoms referred to as **chemical bonds**. The type and extent of that disturbance determines the chemical and physical properties of the new compound. Scientists presently recognize two types of chemical bonds, **ionic bonds** and **covalent bonds**. Ionic bonds result from the transfer of electrons from one atom to another; covalent bonds result from the sharing of electrons between atoms. These two types represent the two extreme kinds of chemical bonds. Many compounds exhibit characteristics consistent with a composite of these two types. The explanation of why and how an element forms a chemical bond will begin with a study of the noble gases.

chemical bond
ionic bond
covalent bond
positive ion
ionization
oxidized
reduced
oxidation state
ionic compound
ionic equation
spectator ion
alkali metal
alkaline earth element
halogen
covalent bonding
molecular orbital
double bond
triple bond
simple ion
complex ion

significance of the noble gases

As mentioned in Chapter 7, the elements in the far right column of the periodic table are unusually stable. Until recently it was believed that they would not form compounds of any kind. In 1963, however, the first noble gas compounds were made by Niel Bartlett in Canada. Since then, several noble gas compounds have been prepared and studied. All have been found to be unstable, and some are explosive in their decomposition.

The electron configurations of the noble gases apparently are unusually stable compared with those of other elements. A closer look at the electron configuration of the noble gases shows that, in every case, the orbitals of each energy level contain the maximum numbers of electrons. This principle is illustrated in Table 8-1.

The almost complete lack of reactivity exhibited by the noble gases has led to several interesting uses of these materials. Valuable documents such as the Constitution of the United States have been encased in glass, with noble gas displacing the normal atmosphere. This inert atmosphere prevents deterioration of the paper by air oxidation.

Old electric light bulbs were evacuated to prevent oxidation of the tungsten filament. Unfortunately, hot tungsten slowly evaporates under these conditions, leaving a black deposit on the inside of the bulb. Today, light bulbs usually are filled with argon, which effectively prevents oxidation and retards evaporation of the filament.

Colorless crystals of xenon tetrafluoride (XeF_4). Chemists in 1962 succeeded in making xenon combine chemically with one other element (fluorine) to form the first binary compound of one of the noble gases. (*Courtesy Argonne National Laboratory.*)

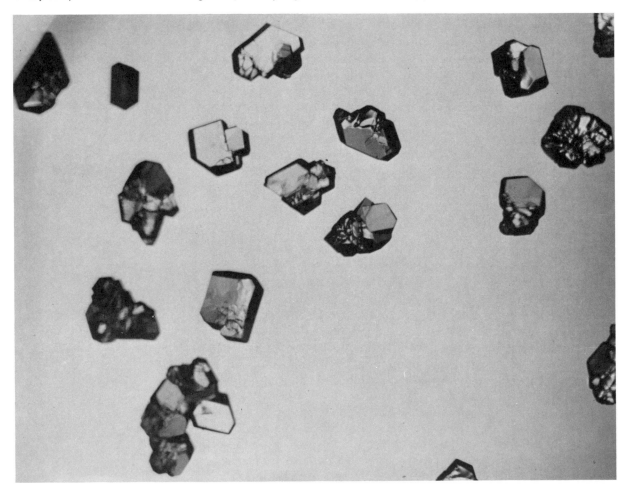

Table 8-1. Electron configuration of the noble gases

He	$1s^2$
Ne	$1s^2 2s^2 2p^6$
Ar	$1s^2 2s^2 2p^6 3s^2 3p^6$
Kr	$1s^2 2s^2 2p^6 3s^2 3p^6 4s^2 3d^{10} 4p^6$
Xe	$1s^2 2s^2 2p^6 3s^2 3p^6 4s^2 3d^{10} 4p^6 5s^2 4d^{10} 5p^6$
Rn	$1s^2 2s^2 2p^6 3s^2 3p^6 4s^2 3d^{10} 4p^6 5s^2 4d^{10} 5p^6 6s^2 4f^{14} 5d^{10} 6p^6$

It may be observed that each noble gas except helium contains a filled outer p sublevel. Since the electron configurations of the noble gases are so stable, it might be expected that other elements would gain or lose electrons in an effort to assume one of these configurations. This actually does occur, and leads to ionic bonding.

ionic bonding

Study of a sodium atom suggests that if the atom were to lose one electron, the remaining electrons would have a configuration resembling that of neon.

$$1s^2 2s^2 2p^6 3s^1 \rightarrow 1s^2 2s^2 2p^6 \ + \ e^{-1}$$

Sodium atom   Sodium ion   Electron
                  (+1 charge)

In the process, the sodium atom would become positively charged, since the number of protons (11) would remain unchanged, but there is now one less electron (10). In this case, the sodium atom would become a **positive ion**. Remember from Chapter 3 that an atom or group of atoms that is charged electrically is called an ion. It can now be seen that the charge on an ion results from the gain or loss of electrons from an atom or group of atoms. The process of forming an ion is called **ionization**. The net result of this process is the formation of a more stable chemical species.

To reach a noble gas configuration, a chlorine atom may lose seven electrons and thus attain the electron configuration of neon. The resulting ion is called a *chloride ion*. The suffix *ide* normally is used to identify simple negatively charged ions.

$$[\text{Ne}]3s^2 3p^5 \rightarrow [\text{Ne}] \ + \ 7\,e^{-1}$$

Chlorine   Chloride ion
atom   (+7 charge)

Unfortunately, the loss of so many electrons normally leads to a less stable particle. Therefore, electron losses of this magnitude normally are not observed. Chlorine can reach a stable, noble gas configuration, however, if instead it gains one electron to assume the argon configuration.

$$[Ne]3s^23p^5 + e^{-1} \rightarrow [Ne]3s^23p^6$$

| Chlorine | Electron | Chloride ion |
| atom | | (−1 charge) |

Electrons lost by one reactant in a chemical reaction are gained by another. For example, if a sodium atom and a chlorine atom chemically interact, the electron lost by the sodium atom will be gained by the chlorine atom. This reaction leads to the formation of sodium chloride, an ionic compound composed of sodium and chloride ions. The reaction can be illustrated by the following chemical equation.

$$Na + Cl \rightarrow Na^{+1} + Cl^{-1}$$

The sodium ion bears a superscript of $+1$ to indicate that it has lost one electron and has a postive charge of one. Similarly, the chloride ion bears a superscript of -1 to indicate that it has gained one electron and has a negative charge of one. In terms of electron configuration, the equation may be written as indicated below.

$$1s^22s^22p^63s^1 + 1s^22s^22p^63s^23p^5 \rightarrow 1s^22s^22p^6 + 1s^22s^22p^63s^23p^6$$

| Sodium atom | Chlorine atom | Sodium ion | Chloride ion |

or:
$$[Ne]3s^1 + [Ne]3s^23p^5 \rightarrow [Ne] + [Ar]$$

| Sodium | Chlorine | Sodium | Chloride |
| atom | atom | ion | ion |

In general, when a metal undergoes a chemical reaction with a nonmetal, electrons are lost by the metal and gained by the nonmetal. When an element loses electrons, it is said to be **oxidized**; when it gains electrons, it is said to be **reduced**. In the equations above, the sodium atom is oxidized and the chlorine atom is reduced. The resulting sodium ion has an **oxidation state** of positive one; the chlorine ion has an oxidation state of negative one.

When elements react, it is not necessary that they do so in a 1 to 1 atomic ratio. Calcium, for example, has the electron configuration $[Ar]4s^2$, and bromine has the electron configuration $[Ar]4s^23d^{10}4p^5$. Calcium needs to lose two electrons to achieve the argon configuration, and bromine needs to gain one electron to achieve the krypton configuration, $[Ar]4s^23d^{10}4p^6$. Thus one calcium atom will react with two bromine atoms.

$$[Ar]4s^2 + 2 ([Ar]4s^23d^{10}4p^5) \rightarrow [Ar] + 2 ([Ar]4s^23d^{10}4p^6)$$

| Calcium atom | 2 Bromine atoms | Calcium ion (+2 charge) | 2 Bromide ions (−1 charge) |

or:
$$Ca + 2 Br \rightarrow Ca^{+2} + 2 Br^{-1}$$

ionic compounds

Compounds formed by the transfer of one or more electrons from one atom to one or more other atoms are called ionic compounds. The positive charges always are balanced by an equal number of negative charges, so the ionic compound has no net charge. When dissolved in water, ionic compounds dissociate into separate ions that exist independently in solution. When the water is evaporated, the ions recombine to form crystals of the original compound.

The same dissociation takes place when two or more ionic compounds are dissolved in the same solution. In this case all the ions exist independently. When the solvent is evaporated, the ions can recombine to form several different compounds. For example, when sodium chloride and calcium hydroxide solutions are added together, the sodium, chloride, calcium, and hydroxide ions present in solution can recombine, on evaporation of the solvent, to form sodium hydroxide, sodium chloride, calcium hydroxide, and calcium chloride.

With some combinations, two ions may form a combination that is insoluble in water. In this case, the insoluble combination of ions will precipitate from the solution, leaving the other ions in solution. For example, if a solution of sodium chloride in water is mixed with a solution of silver nitrate, a white precipitate of silver chloride is formed. The equation for this process may be written as follows.

$$AgNO_3 + NaCl \rightarrow \underline{AgCl} + NaNO_3$$

Silver chloride is underlined in this equation because it forms as a precipitate. Since only the silver and chloride ions participate in the chemical reaction leading to silver chloride, it is also possible to write the equation using only these materials.

$$Ag^{+1} + Cl^{-1} \rightarrow \underline{AgCl}$$

This type of equation is known as an ionic equation because it only includes the ions that actually participate in the chemical reaction. An ionic equation recognizes that it does not matter what compound containing silver ions is mixed with what compound containing chloride ions. Silver chloride will always precipitate. Since the other ions that make up the original silver compound and the original chloride compound do not participate in the reaction, they are sometimes called spectator ions. A compound of the spectator ions could be obtained by evaporating the solvent. For example, in the sodium chloride–silver nitrate mixture, sodium nitrate would be formed on evaporation of the solvent.

Other examples of ionic equations are given below.

$$Ca^{+2} + CO_3^{-2} \rightarrow CaCO_3$$

$$2\ Ag^{+1} + SO_4^{-2} \rightarrow Ag_2SO_4$$

$$Al^{+3} + 3\ OH^{-1} \rightarrow Al(OH)_3$$

prediction of ionic charge

Prediction of the principal oxidation states or charges of the elements is another of the successes of the periodic table. A further look at the periodic table shows that each element in the far left column needs to lose just one electron to attain the electron configuration of the noble gas that precedes it in atomic number. Thus all these elements tend to have a positive one oxidation state in compounds. These Group 1A elements are highly reactive. They are not found free in nature and even will react vigorously with water. They commonly are called the alkali metals because they form strong alkalis or bases with water. Acids and bases will be discussed in Chapter 11.

The second column from the left in the periodic table contains metals that must lose two electrons each to attain a noble gas configuration. These Group IIA elements are known as the alkaline earth elements. They are also highly reactive and form strong bases with water, but they are not as reactive as the alkali metals.

The transition elements (Group IB-VIIB, VIII) represent a special situation. These elements do not always follow the rules described earlier, because of their partially filled orbitals. They tend to exhibit more than one oxidation state. Iron, for example, commonly is found in the positive two (ferrous) and positive three (ferric) oxidation states. The explanations for these particular oxidation states are quite complicated and may be found in more advanced books.

Left: Aluminum ingots. *Right*: A gallium crystal. Both elements belong to Group IIIA of the periodic table. (*Courtesy left, Kaiser Aluminum and Chemical Company; right, Aluminum Company of America.*)

The Group IIIA elements tend to lose three electrons each. This gives them an oxidation state of a positive three. They are not nearly so active as the alkali or alkaline earth metals, however.

The Group VIIA elements need only one electron each to attain the next highest noble gas configuration. These elements are known as the halogens and normally exist in the negative one oxidation state in compounds. They are highly reactive and are not found naturally as free elements.

The Group VIA elements normally exist in the negative two oxidation state, and Group VA elements usually exist in the negative three oxidation state in compounds. These elements are less reactive than the halogens, and some occur in elemental form in nature (usually as molecules containing two or more like atoms). Some of the heavier Group VA elements occasionally are found in the positive five oxidation state.

The Group IVA elements exhibit both the positive four and negative four oxidation states. More commonly, these elements exhibit a different type of chemical bonding known as covalent bonding

covalent bonding

Another common situation observed in compound formation is the tendency for two or more atoms to combine so that neither atom really has gained or lost electrons. In effect, both atoms *share* electrons to achieve stability. This type of chemical bond is called a *covalent bond*.

The hydrogen molecule is a typical example of a covalently bonded unit of matter. Hydrogen gas exists almost exclusively in the form of diatomic molecules, H_2.

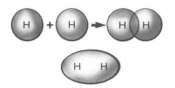

When two hydrogen atoms approach close enough to form a bond, the 1s orbitals of the atoms overlap as shown at the left.

When this happens, the two atomic orbitals reorganize so that they form a bonding molecular orbital, again as shown at the left.

The molecular orbital actually is the covalent bond holding the two atoms together in the hydrogen molecule. In general, all covalent bonds are composed of molecular orbitals formed by the overlap of an orbital of one atom with an orbital of another.

The halogens also tend to exist predominantly as diatomic molecules. Consider chlorine as an example. Each chlorine atom contains 17 electrons. Fortunately, only the outer, or higher-energy, electrons are involved in bonding. From the position of chlorine in the periodic table, it can be seen that the outer electrons of chlorine are in 3s and 3p orbitals. This can be verified by writing the complete electron configuration as $[Ne]3s^23p^5$. Actually, all these orbitals are filled except one 3p orbital that is only half-filled (contains only one electron).

A simple picture of the chlorine molecule can be prepared by imagining the formation of a molecular orbital between the two half-filled p orbitals of two chlorine atoms, much like the hydrogen molecule.

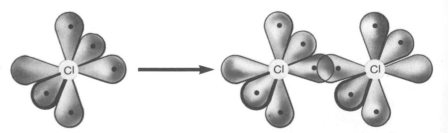

This representation is somewhat awkward. For this reason, orbitals normally are not drawn as shown. Instead, electrons in the outer energy level are represented by small dots or x's, which are written around the symbol of the element. When two electrons occupy one orbital, they are indicated as an electron pair. Using this convention, the previous equation is pictured accordingly as at the right.

$$:\ddot{\underset{..}{Cl}}\cdot \; + \; \cdot\ddot{\underset{..}{Cl}}: \; \rightarrow \; :\ddot{\underset{..}{Cl}}:\ddot{\underset{..}{Cl}}:$$

$$Cl \; + \; Cl \; \rightarrow \; Cl_2$$

This symbolism makes it possible to represent easily many atoms and molecules, as shown in the examples at the right.

In the methane molecule, x's indicate the electrons from one atom and dots distinguish the electrons from the other atom.

Sometimes more than one orbital from an atom is implicated in covalent bonding with another atom. This leads to involvement of multiple electron pairs and consequently to multiple covalent bonds. If two electron pairs are involved, a double bond is formed.

$$:\ddot{\underset{..}{F}}\cdot \qquad \cdot\dot{\underset{.}{N}}\cdot \qquad :\dot{\underset{..}{O}}:$$

$$\begin{array}{c} H \\ \overset{\times\cdot}{H\!:\!\underset{\cdot\times}{C}\!\overset{\times}{:}\!H} \\ H \end{array} \qquad :\ddot{\underset{..}{I}}:\ddot{\underset{..}{I}}:$$

Methane   Iodine

$$\dot{\underset{.}{O}}:\overset{\times}{C}\overset{\times}{:}\dot{\underset{.}{O}}$$

Carbon dioxide

$$\begin{array}{c} H \\ \overset{+}{C}\overset{\times}{:}\dot{\underset{.}{O}}: \\ H \end{array}$$

Formaldehyde

Triple bonds such as the following are also known.

$$:N\overset{..}{::}N:$$

Nitrogen

$$H\overset{\times}{:}C\overset{\times\times}{\underset{\times\times}{::}}C\overset{\times}{:}H$$

Acetylene

Often the two electrons that are involved in the formation of a covalent bond are represented simply by a straight line connecting the two atoms being bonded together, as shown at the right.

When this symbolism is used, the extra electrons that are not involved in bonding usually are omitted entirely.

$$:\ddot{\underset{..}{Cl}}\cdot \; + \; \cdot\ddot{\underset{..}{Cl}}: \; \rightarrow \; :\ddot{\underset{..}{Cl}}\!-\!\ddot{\underset{..}{Cl}}:$$

$$Cl + Cl \rightarrow Cl\!-\!Cl$$

Other illustrations are given below.

$$N\equiv N$$

Nitrogen

$$\begin{array}{c} H \\ \diagdown \\ \quad\;\; C\!=\!O \\ \diagup \\ H \end{array}$$

Formaldehyde

$$H\!-\!C\equiv C\!-\!H$$

Acetylene

$$O\!=\!C\!=\!O$$

Carbon dioxide

$$C\equiv O$$

Carbon monoxide

$$F\!-\!F$$

Fluorine

orientation of atoms in space

The orbital concept of the atom can predict the spatial orientation of atoms in a molecule. Consider the water molecule as an example. Water has the formula H_2O. The oxygen atom has the electron distribution $[He]2s^2 2p^4$. This means that the $2s$ and one of the $2p$ orbitals are filled and two of the $2p$ orbitals contain one electron each. If the p orbitals are drawn, it is seen that the two half-filled orbitals are at right angles to one another.

If two hydrogen atoms now are allowed to form covalent bonds with the two half-filled p orbitals of oxygen, the H—O—H bond angle will be predicted to be 90°.

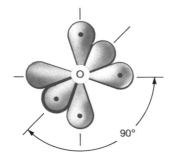

The actual observed angle is 104°. The slightly larger observed bond angle may be due to repulsion between the two hydrogen nuclei, which are both positively charged.

complex ions

Previous discussions have considered only simple ions formed by the gain or loss of electrons by individual atoms. Ions also can be formed from groups of atoms. These are called complex ions. The hydroxide ion (OH^-) is a typical example of this type of ion. As already shown, oxygen needs two electrons to achieve the stable neon configuration. It also has been shown that oxygen can form covalent bonds with hydrogen to form water. Oxygen makes the hydroxide ion by forming one covalent bond with a hydrogen atom and gaining one additional electron from some other source, such as a sodium atom, as seen at the left.

As a consequence, the hydroxide ion has a negative one charge.

The ammonium ion is formed when a hydrogen ion reacts to make a covalent bond with the extra pair of electrons of the ammonia molecule, as shown at the top of the following page.

$$\ddot{\underset{\cdot\cdot}{O}}{\overset{\times}{}}H \equiv OH^{-1}$$

$$H-\underset{\underset{H}{|}}{\overset{\overset{H}{|}}{N}}: + H^{+1} \rightarrow H-\underset{\underset{H}{|}}{\overset{\overset{H}{|}}{N}}-H \quad {}^{+1}$$

A number of ions of the complex type are known. Table 8-2 gives the formulas and names of some common complex ions.

Table 8-2. Some common complex ions

Formula	Name
OH^{-1}	hydroxide ion
$NO_3{}^{-1}$	nitrate ion
$NO_2{}^{-1}$	nitrite ion
$SO_3{}^{0}$	sulfito ion
$SO_4{}^{-2}$	sulfate ion
$PO_4{}^{-3}$	phosphate ion
$NH_4{}^{+1}$	ammonium ion
$CH_3{}^{+1}$	methonium ion
H_3O^{+1}	hydronium ion
$CO_3{}^{-2}$	carbonate ion

In chemical formulas for compounds, complex ions normally are indicated as separate, distinctive units.

$$NaNO_3 \quad KOH \quad NH_4Cl$$

If two or more complex ion groups are needed in a formula, they are enclosed in parentheses with a subscript to indicate the number of groups involved.

$$(NH_4)_2S \quad Mg(NO_3)_2 \quad Al_2(SO_4)_3$$

Such a designation gives more information about the composition of a compound than the simpler designations that follow.

$$N_2H_8S \quad MgN_2O_6 \quad Al_2S_3O_{12}$$

Although the chemical bonding within a complex ion is covalent, the "external" bonding usually is ionic, as indicated here. Thus, it is possible for a compound to contain both ionic and covalent bonds.

$$Na^{+1}OH^{-1}$$

ionic bond ⟋ ⟍ covalent bond

study questions

1 What is unique about the electron configurations of the noble gases?

2 Why does an atom react to form a chemical bond?

3 What determines whether an element will tend to gain or to lose electrons in a chemical reaction?

4 Distinguish between an ionic bond and a covalent bond.

5 What is the difference between a double covalent bond and a single covalent bond?

6 How does an ion compare with its corresponding atom in terms of nuclear composition and electronic structure?

7 Why does an atom that loses electrons become positively charged?

8 Write electron dot diagrams showing the outer electronic structure for each of the following elements.
a. Be   c. Cl   e. B
b. Mg   d. P

9 In terms of electron configurations, write an equation to show the conversion of each of the following atoms to the corresponding ion.
a. Li   b. O   c. Mg

10 Predict the charges of ions that would be obtained from the following atoms.
a. Ba   c. I   e. N
b. S   d. K   f. Al

11 In an ordinary chemical reaction, would cesium form +2 ions? Would it form −1 ions? Why?

12 Predict formulas for chemical compounds that would form by reactions between the following elements.
a. K and Cl   d. Al and Br
b. Ca and I   e. Mg and S
c. Na and O

13 In the reactions in Question 12, which elements become oxidized and which become reduced?

14 Write formulas for the following compounds.
a. Barium sulfate   e. Lithium carbonate
b. Silver phosphate   f. Aluminum nitrate
c. Magnesium hydroxide

special problems

1 Represent the following covalent molecules by electron dot diagrams like those shown for methane, carbon dioxide, and so forth, on page 129.

a. NH_3 b. $CHCl_3$ c. H_2O_2 d. CS_2

2 Complete and balance the following equations.

a. $KCl + AgNO_3 \rightarrow$

b. $HCl + MgSO_4 \rightarrow$

c. $Ca(OH)_2 + HBr \rightarrow$

d. $FeBr_3 + K_2CO_3 \rightarrow$

e. $BaO + HNO_3 \rightarrow$

3 Write ionic equations to represent the formation of the following compounds from the appropriate ions.

a. Silver bromide

b. Barium sulfate

c. Magnesium carbonate

d. Aluminum hydroxide

chapter **9**

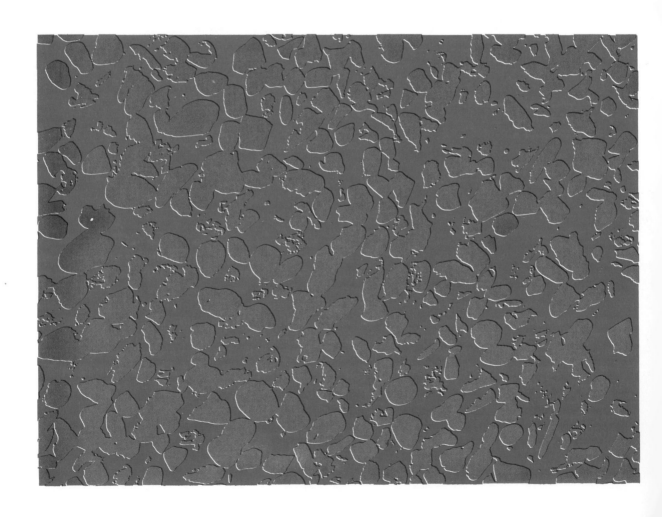

quantitative relationships

Chemistry is a quantitative science. Consequently, it is possible to predict how much of a particular product will be formed from a known amount of starting material. Conversely, it is possible to calculate how much starting material is required to form a desired quantity of product. Two chemical laws govern the quantitative relationships of chemistry, the Law of Conservation of Mass and the Law of Constant Composition. Both these laws were discussed briefly in Chapter 1.

formula weight
molecular weight
gram formula weight
mole

molecular weight

The concepts of atomic weight and the gram-atom were discussed in Chapter 4. These are especially useful concepts when working with atoms and elements. When working with molecules and compounds, however, a new, more general concept is needed. This new concept is the **formula weight** or **molecular weight**. The formula weight or molecular weight of a compound is the sum of the atomic weights of its components as represented in its formula. Normally these two terms are used interchangeably, although, strictly speaking, *molecular weight* applies only to covalent compounds or those whose particles of matter are molecules. The more general term *formula weight* applies to both ionic and covalent compounds.

According to the definition given, the formula weight of sodium chloride (NaCl) is 58.5 amu. This is due to the fact that the atomic weight of sodium is 23 amu and that of chlorine is 35.5 amu. Several additional examples follow.

This bag of fertilizer-grade ammonium nitrate (NH_4NO_3) shows an analysis of 33.5% nitrogen. Pure ammonium nitrate contains 35% nitrogen, as determined by dividing the weight of nitrogen (28) in ammonium nitrate by the formula weight (80) of ammonium nitrate and multiplying by 100. Obviously some other nonnitrogenous components are present in this grade of ammonium nitrate. (*Courtesy Collier Carbon and Chemical Corporation, a wholly-owned subsidiary of Union Oil Company of California*)

1 Formula weight of calcium carbonate (limestone) ($CaCO_3$)

$$
\begin{array}{llll}
\text{Ca} & 40 \times 1 = & 40 \text{ amu} \\
\text{C} & 12 \times 1 = & 12 \text{ amu} \\
\text{O} & 16 \times 3 = & \underline{48 \text{ amu}} \\
& & 100 \text{ amu}
\end{array}
$$

2 Formula weight of sulfuric acid (H_2SO_4)

$$
\begin{array}{llll}
\text{H} & 1 \times 2 = & 2 \text{ amu} \\
\text{S} & 32 \times 1 = & 32 \text{ amu} \\
\text{O} & 16 \times 4 = & \underline{64 \text{ amu}} \\
& & 98 \text{ amu}
\end{array}
$$

3 Formula weight of aluminum sulfate ($Al_2(SO_4)_3$)

$$
\begin{array}{llll}
\text{Al} & 27 \times 2 = & 54 \text{ amu} \\
\text{S} & 32 \times 3 = & 96 \text{ amu} \\
\text{O} & 16 \times 12 = & \underline{192 \text{ amu}} \\
& & 342 \text{ amu}
\end{array}
$$

gram formula weight (gfw) or mole

The gram formula weight of a compound is its molecular weight or formula weight expressed in grams. Like the gram atomic weight, which represents 6.02×10^{23} *atoms*, the gram formula weight represents 6.02×10^{23} *molecules*. The gram formula weight commonly is called a mole. Stated somewhat differently, a mole represents the weight of 6.02×10^{23} molecules of a compound.

It is often desirable to know how many moles are contained in a given quantity of a compound. This can be determined easily from the following relationship.

$$
\frac{\text{weight in grams}}{\text{gram formula weight}} = \text{number of moles}
$$

For example, iron oxide (Fe_2O_3) is the major component of iron ore and common rust. To determine the number of moles of iron oxide in 480 grams of this compound, the following procedure may be used.

1 Determine the formula weight of iron oxide.

$$
\begin{array}{llll}
\text{Fe} & 56 \times 2 = & 112 \text{ amu} \\
\text{O} & 16 \times 3 = & \underline{48 \text{ amu}} \\
& & 160 \text{ amu}
\end{array}
$$

2 Using the expression given in the discussion above,

$$\frac{weight}{GFW} = \frac{480 \text{ g}}{160 \text{ g/mole}} = 3 \text{ moles}$$

It is also useful to be able to calculate the number of grams of a substance in a given number of moles. For example, the active ingredient in many oven cleaners is lye (sodium hydroxide, NaOH). How many grams of sodium hydroxide are in 1.5 moles of this compound?

1 Determine the formula weight of sodium hydroxide.

$$
\begin{array}{lll}
\text{Na} & 23 \times 1 = & 23 \text{ amu} \\
\text{O} & 16 \times 1 = & 16 \text{ amu} \\
\text{H} & 1 \times 1 = & \underline{1 \text{ amu}} \\
& & 40 \text{ amu}
\end{array}
$$

2 Using the expression given earlier,

$$\frac{weight}{GFW} = moles$$

$$weight = GFW \times moles$$

$$weight = 40 \text{ g/mole} \times 1.5 \text{ moles}$$

$$weight = 60 \text{ grams}$$

utility of the mole

The mole concept is exceedingly useful because it allows a chemical reaction to be expressed in terms of molecules rather than weight. For example, a typical chemical reaction is the combustion of natural gas, methane (CH_4), in air. The chemical equation for this reaction is as follows.

$$CH_4 + 2 O_2 \rightarrow CO_2 + 2 H_2O$$

The equation indicates that one molecule of methane will interact with two molecules of oxygen to produce one molecule of carbon dioxide and two molecules of water. The equation also shows directly, by its coefficients, that one mole of methane will be consumed by two moles of oxygen to produce one mole of carbon dioxide and two moles of water.

Although the relationships between starting materials and products in an equation are given in moles (or molecules), all necessary information is available for determining weight relationships between these substances as well. These weight relationships may be calculated in the following manner for the combustion of one mole (16 g) of methane.

$$CH_4 \quad + \quad 2\,O_2 \quad \rightarrow \quad CO_2 \quad + \quad 2\,H_2O$$

GFW   16 g   32 g   44 g   18 g

Multiply by number of moles involved (coefficient from equation)

$$\times\,1 \qquad \times\,2 \qquad \times\,1 \qquad \times\,2$$

$$16\text{ g }CH_4 + 64\text{ g }O_2 \rightarrow 44\text{ g }CO_2 + 36\text{ g }H_2O$$

Thus 16 grams of methane are consumed by 64 grams of oxygen and produce 44 grams of carbon dioxide and 36 grams of water. Because this is a balanced equation, the total mass on each side of the equation is identical (80 grams for this particular equation).

The weight relationships between starting materials and products are constant for a given reaction and are represented by the balanced chemical equation. Therefore, the same amount of products will be formed even though one of the reactants is present in excess. For example, in the combustion of methane in air, oxygen always is

Modern instrumentation such as this differential refractometer permits a scientist to determine experimentally the molecular weight of a substance. (*Courtesy E. I. du Pont de Nemours & Co., Wilmington, Delaware*)

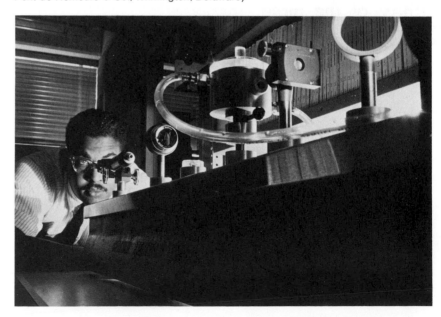

present in large excess. Nevertheless, one mole (16 grams) of methane will react with only two moles (32 grams) of oxygen to produce only one mole (44 grams) of carbon dioxide and two moles (36 grams) of water, regardless of the large quantities of oxygen in the air.

The relationships expressed by a balanced chemical equation can determine the amounts of products or additional starting materials required in a chemical reaction. The following examples illustrate this point.

1 When vinegar (a dilute solution of acetic acid, $HC_2H_3O_2$, obtained by fermentation) is mixed with baking soda (sodium hydrogencarbonate, $NaHCO_3$), the solution fizzes, owing to formation of carbon dioxide. The reaction is represented by the following equation.

$$HC_2H_3O_2 + NaHCO_3 \rightarrow NaC_2H_3O_2 + H_2O + CO_2$$

How many moles of sodium hydrogencarbonate are required to react with 3 moles of acetic acid?

According to the balanced equation, the molar ratio of sodium hydrogencarbonate to acetic acid is 1 to 1.

$$\text{molar ratio} = \frac{1 \text{ mole } NaHCO_3}{1 \text{ mole } HC_2H_3O_2}$$

If this ratio is multiplied by the moles of acetic acid that are given, the units will cancel, giving the correct answer.

$$\frac{1 \text{ mole } NaHCO_3}{1 \text{ mole } HC_2H_3O_2} \times 3 \text{ moles } HC_2H_3O_2 = 3 \text{ moles } NaHCO_3$$

2 Ammonia is manufactured in huge fertilizer plants by the direct combination of hydrogen and nitrogen. Ammonia is used as a fertilizer, either directly or as a related compound such as ammonium nitrate or urea.

How many moles of ammonia are produced when 6 moles of hydrogen react with nitrogen (in excess) as follows.

$$N_2 + 3 H_2 \rightarrow 2 NH_3$$

From the balanced equation, it can be determined that there are 2 moles of NH_3 produced for every 3 moles of hydrogen consumed. The molar ratio of ammonia to hydrogen is listed accordingly.

$$\frac{2 \text{ moles } NH_3}{3 \text{ moles } H_2} = \text{molar ratio}$$

Multiplying by the moles of hydrogen available gives the correct answer.

$$\frac{2 \text{ moles } NH_3}{3 \text{ moles } H_2} \times 6 \text{ moles } H_2 = 4 \text{ moles } NH_3$$

Most commonly, the data for these problems are given in grams rather than in moles. In some cases they might even be given in pounds or ounces. Since chemical equations give molar relationships directly, not weight relationships, it is necessary first to convert given weights to molar quantities. Once this is accomplished, the problem can be solved as before.

1 Butane is used for cooking and heating in some rural areas. How many grams of carbon dioxide are produced by the combustion of 11.6 grams of butane, according to the following equation?

$$2\ C_4H_{10} + 13\ O_2 \rightarrow 8\ CO_2 + 10\ H_2O$$

a. The given weight of butane must be converted to *moles* of butane.

$$moles = \frac{weight}{gram\ formula\ weight} = \frac{11.6\ g\ butane}{58\ g\ butane/mole}$$

$$= 0.2\ mole$$

The formula weight was determined by adding the atomic weights of all component atoms shown in the correct formula, as discussed earlier.

b. The molar ratio of CO_2 to C_4H_{10} is

$$\frac{8\ moles\ CO_2}{2\ moles\ C_4H_{10}} = molar\ ratio$$

Therefore

$$\frac{8\ moles\ CO_2}{2\ moles\ C_4H_{10}} \times 0.2\ mole\ C_4H_{10}$$

$$= 0.8\ mole\ CO_2$$

c. Since the problem concerns the number of *grams* of CO_2 produced, the answer calculated above (0.8 moles) now must be converted to grams.

$$moles\ CO_2 = \frac{wt.\ CO_2}{g.\ formula\ wt.\ CO_2}$$

$$moles\ CO_2 \times g.\ formula\ wt.\ CO_2 = wt.\ CO_2$$

Substituting appropriate values into this equation provides the final answer.

$$0.8\ mole\ CO_2 \times 44\ g/mole = wt.\ CO_2$$

$$wt.\ CO_2 = 35.2\ grams$$

2 As noted in Chapter 20, an average man produces 1 kilogram of carbon dioxide per day. Lithium hydroxide removes carbon dioxide from the cabin of a spacecraft by the reaction shown below.

$$2\ LiOH + CO_2 \rightarrow Li_2CO_3 + H_2O$$

How many kilograms of lithium hydroxide must be provided in the cabin for each day one man spends in space?

a. From the molecular weight of carbon dioxide, the number of moles of carbon dioxide in 1 kilogram may be determined.

$$\frac{1000 \text{ g}}{44 \text{ g/mole}} = 22.8 \text{ moles}$$

b. From the balanced chemical equation, the molar ratio may be determined.

$$\frac{2 \text{ moles LiOH}}{1 \text{ mole CO}_2} = \text{molar ratio}$$

c. Multiplying the molar ratio by the number of moles of carbon dioxide to be absorbed gives the moles of lithium hydroxide required to remove 1 kilogram of carbon dioxide.

$$\frac{2 \text{ moles LiOH}}{1 \text{ mole CO}_2} \times 22.8 \text{ moles CO}_2 = 45.6 \text{ moles LiOH}$$

d. The number of grams of lithium hydroxide required is determined by converting the moles of lithium hydroxide to grams.

$$\text{moles LiOH} = \frac{\text{wt. LiOH}}{\text{g. formula wt. LiOH}}$$

wt. LiOH	= moles LiOH × g. formula wt. LiOH
wt. LiOH	= 45.6 moles × 24 g/mole
wt. LiOH	= 1090 g
wt. LiOH	= 1.1 kilograms

study questions

1 State the Law of Conservation of Mass. (See Chapter 3 if necessary.)

2 State the Law of Constant Composition. (See Chapter 3 if necessary.)

3 What importance do the Laws of Conservation of Mass and Constant Composition have in chemistry?

4 Why is chemistry a quantitative science?

5 What practical applications can you name for the quantitative concepts of chemistry?

6 Determine the formula weight of each of the following compounds.

    a. CaI_2 d. $Mg(NO_3)_2$

    b. K_2O e. $BaSO_4$

    c. Al_2S_3 f. $(NH_4)_2Cr_2O_7$

7 How many grams of compound are contained in one mole of each of the compounds in Question 6?

special problems

1 How many grams of compound are contained in each of the following?

    a. 2 moles of $CaCO_3$ c. 0.3 mole of C_6H_6

    b. 5.7 moles of HNO_3 d. 0.008 mole of $FeCl_3$

2 To how many moles of each substance are the following quantities of material equivalent?

    a. 36 grams of H_2O c. 300 grams of Na_2SO_4

    b. 22 grams of CO_2 d. 50 grams of H_3PO_4

3 In the following reaction:

$$Na_2CO_3 + 2\ HCl \rightarrow 2\ NaCl + H_2O + CO_2$$

    a. How many moles of Na_2CO_3 would be required to react with 3 moles of HCl? With 0.5 mole of HCl?

    b. How many moles of Na_2CO_3 would be required to produce 5 moles of CO_2? 1 mole of NaCl? (Assume excess HCl is present in each case).

4 Suppose that 190 grams of $MgCl_2$ are treated with an excess of $AgNO_3$ in the following reaction.

$$MgCl_2 + 2\ AgNO_3 \rightarrow 2\ AgCl + Mg(NO_3)_2$$

How many grams of AgCl would be produced?

To solve this problem, answer the following questions.

    a. How many moles of $MgCl_2$ are used in the reaction?

    b. How many moles of AgCl would be produced per mole of $MgCl_2$ used?

    c. Based on your answers to (a) and (b), how many moles of AgCl would be produced in this reaction from the given amount of $MgCl_2$?

    d. How many grams of AgCl are equivalent to the number of moles of AgCl shown in your answer to (c)?

5 If 588 grams of sulfuric acid were to react with excess iron metal in the following reaction, how many grams of iron sulfate would be produced?

$$2 \text{ Fe} + 3 \text{ H}_2\text{SO}_4 \rightarrow \text{Fe}_2(\text{SO}_4)_3 + 3 \text{ H}_2$$

6 What is the percentage composition by weight of each element in the following compounds?
a. P_2O_5 b. $ZnCO_3$ c. $(NH_4)_2S$

chapter **10**

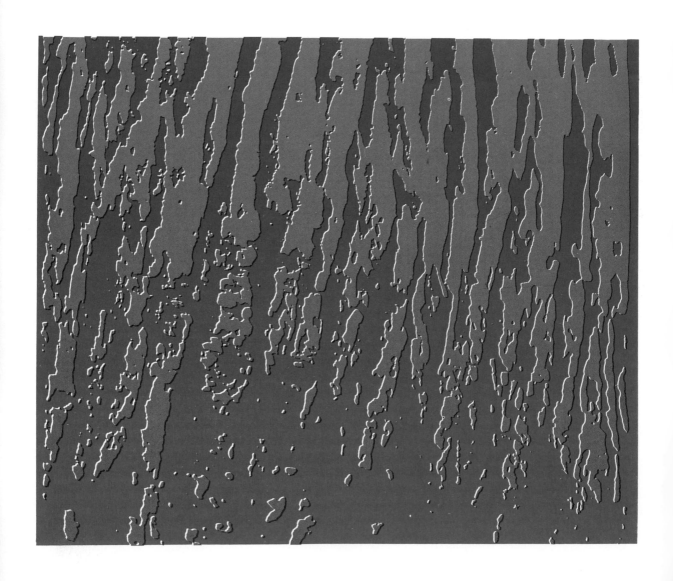

oxidation-reduction

A particularly important class of chemical reactions involves the transfer of electrons from one chemical species to another. Reactions of this type are called oxidation-reduction reactions, or sometimes redox reactions. Redox reactions are involved in such phenomena as iron rusting, wood combustion, the electroplating industry, and functioning of automobile batteries. Knowledge of redox reactions is important for understanding many natural phenomena and industrial processes.

oxidation-reduction reaction
redox reaction
oxidation
oxidized
reduced
reduction
oxidizing agent
reducing agent
oxidation number
half reaction
oxidation potential
electrolyte
voltaic cell

basic concepts

The rusting of iron is an oxidation-reduction reaction.

$$4 \, Fe + 3 \, O_2 \rightarrow 2 \, Fe_2O_3$$

In an oxidation-reduction reaction, one species (iron in the above example) loses electrons, thus acquiring a more positive electrical charge. This process is called oxidation, and the species that has undergone oxidation is said to have been oxidized. Since all compounds are electrically neutral, there also must be a species in the reaction that gains electrons. Oxygen is that species in the example above. A species that gains electrons is said to be reduced, and the process of gaining electrons is known as reduction. For each oxidation, there must be a corresponding reduction to maintain electrical neutrality.

In the rusting of iron, iron changes from the zero oxidation state of the metal to the +3 oxidation state in iron oxide, or rust. Oxygen changes from the zero oxidation state of elemental oxygen to the −2 oxidation state in iron oxide.

The species that accepts electrons (becomes reduced) in a redox

In making steel (*right*), iron ore (iron oxide, Fe_2O_3) is reduced to free iron by heating with oxygen and carbon, yielding a mixture of carbon monoxide and carbon dioxide as a side product. In this picture, molten iron is being charged into a large, basic-oxygen-process, steel-making furnace where the iron will be alloyed with other materials to make steel. (*Courtesy United States Steel Corporation*)

Rusting of iron (*below*). Iron is being oxidized (losing electrons), and oxygen is being reduced (gaining electrons). (*Photograph by A. Marshall Licht*)

reaction sometimes is referred to as the oxidizing agent, because it is the species that is responsible for the oxidation, or the loss of electrons from another species. Conversely, the species that loses electrons (becomes oxidized) sometimes is called the reducing agent, because it is the species that is responsible for the reduction, or the gain of electrons by another species. In the example discussed earlier, iron is the reducing agent and oxygen is the oxidizing agent.

oxidation number

To understand and correctly analyze redox reactions, it is necessary first to determine oxidation numbers or states for elements as they occur in different substances. The oxidation number of an atom is the apparent or real charge of that atom in a substance. This number is the actual charge of an atom in an ionic species. In covalent species, the oxidation number is the charge an atom would have if it were ionic. Some oxidation numbers may be determined by proper use of the periodic table, as was described in Chapter 8. The following rules can be used to assign oxidation numbers to elements in ionic

Oxidation-reduction reactions are responsible for the corrosion of metals. The rusting of iron is a common example.

$$4\ Fe + 3\ O_2 \rightarrow 2\ Fe_2O_3$$

In this case, iron is oxidized, and oxygen is reduced.

Galvanized iron is iron that has been coated with zinc. Zinc is more easily oxidized (has a higher oxidation potential) than iron. Thus the zinc corrodes and falls off the iron. As long as the zinc coating remains, the iron will not rust.

Aluminum is easily oxidized, but tends to form a thin film of aluminum oxide on its surface. This surface film effectively prevents further corrosion, making aluminum a useful structural material even though it has a relatively high oxidation potential.

Copper has a relatively low oxidation potential and thus is not oxidized easily. It does oxidize slowly on exposure to the weather, however. Copper oxide itself is black, but it absorbs carbon dioxide from the air to form copper carbonate ($CuCO_3$), which is responsible for the green color of copper lettering on buildings.

as well as covalent bonding situations. The theoretical bases for these rules have been discussed earlier.

1 All elements, when free and uncombined with other elements, have oxidation numbers of zero.
2 Hydrogen and oxygen, when involved in compounds, generally have oxidation numbers of $+1$ and -2 respectively.
3 Elements in Families IA, IIA, and IIIA have oxidation numbers of $+1$, $+2$, and $+3$ respectively, when in compounds.
4 The sum of the oxidation numbers of atoms shown in the chemical formula for a compound is zero.
5 The sum of the oxidation numbers of atoms shown in the chemical formula for an ion is equal to the charge of that ion.

The uses of these rules can be illustrated by the determination of the oxidation number of nitrogen in each of the following species.

1 N_2: oxidation number of zero. This is elemental nitrogen (rule 1).
2 NH_3: oxidation number of -3.
The oxidation number of H is $+1$ (rule 2).
The total charge of three hydrogens is $+3$.
Therefore, the oxidation number of N is -3 (rule 4).
3 KNO_2: oxidation number of $+3$.
The oxidation number of K is $+1$ (rule 3).
The oxidation number of O is -2 (rule 2).

The total charge of two oxygens is −4.
Therefore, the oxidation number of N is +3 (rule 4).

$$(+1 + 3 - 4 = 0)$$

4 N^{-3}: oxidation number of −3 (rule 5).
5 NO_3^{-1}: oxidation number of +5.
The oxidation number of O is −2, and the total charge of three oxygens is −6 (rule 2).
Therefore, the oxidation number of N is +5 (rule 5).

$$(+5 - 6 = -1)$$

half reactions

The necessity for both oxidation and reduction to occur in a single reaction permits division of the reaction into two parts, an oxidation part and a reduction part. The two parts are known as half reactions. When representing half reactions by equations, the gain or loss of electrons is represented by the symbol e^- for electrons.

Consider, for example, the violent reaction of sodium metal with water to form sodium hydroxide and hydrogen gas.

$$2\ Na + 2\ H_2O \rightarrow 2\ NaOH + H_2$$

In this reaction, sodium metal with an oxidation state of zero is oxidized to sodium +1 ions with the loss of one electron per atom.

$$Na \rightarrow Na^{+1} + e^{-1}$$

At the same time, one of the hydrogen atoms in the water molecule is reduced from its oxidation state of +1 to free hydrogen with an oxidation state of zero.

$$2\ H_2O + 2\ e^{-1} \rightarrow 2\ OH^{-1} + H_2$$

Equations of half reactions are balanced by adjusting the numbers of atoms of each element on either side of the equation. Electrons then are added to the appropriate side of the equation, to make the charges balance. Thus a half reaction represents a balanced portion of the overall reaction and represents either an oxidation or a reduction. Appropriate treatment and combination of the two half reactions yields the overall redox reaction.

Each half reaction may be considered as proceeding from either direction. The sodium half reaction, for example, may be considered as follows.

$$Na \rightarrow Na^{+1} + e^{-1}$$

or

$$Na^{+1} + e^{-1} \rightarrow Na$$

The top equation represents the oxidation of sodium metal to sodium ions. The bottom equation represents the reduction of sodium ions to sodium metal. A given half reaction, then, can represent either an oxidation or a reduction, depending on the direction in which it is written.

oxidation potential

The direction in which a redox reaction will proceed spontaneously is determined by the relative oxidizing and reducing powers of the starting materials compared with those of the final products. Consider again the reaction of sodium with water.

$$2 \ (Na \rightarrow Na^{+1} + e^{-1})$$
$$\underline{2 \ H_2O + 2 \ e^{-1} \quad \rightarrow 2 \ OH^{-1} + H_2}$$
$$2 \ Na + 2 \ H_2O \quad \rightarrow 2 \ Na^{+1} + 2 \ OH^{-1} + H_2$$

(The overall reaction is shown below the two half reactions.)

The reaction as given above proceeds spontaneously in the direction shown because sodium is a stronger reducing agent than water.

As the result of careful experimentation, a scale of oxidation potential or ability to be oxidized has been developed. The oxidation potential is a quantitative measure of the relative oxidizing characteristics of a half reaction. It is given the symbol E° and is expressed in volts. A representative grouping of some common half reactions and their oxidation potentials is given in Table 10-1.

Table 10-1 can be used in a qualitative manner to predict the outcome of selected oxidation reactions. Some metals, for example, spontaneously will displace other metals from compounds in solution. An iron nail placed in a solution of copper sulfate ($CuSO_4$) will dissolve, leaving a deposit of copper metal. This could be predicted by looking at the oxidation potentials in Table 10-1. Iron has a larger (more positive) oxidation potential than copper.

Table 10-1. Some standard oxidation potentials at 25°C

Half Reaction	Oxidation Potential (E°)
$K \rightarrow K^{+1} + e^-$	2.93 v
$Ca \rightarrow Ca^{+2} + 2\ e^-$	2.87 v
$Na \rightarrow Na^{+1} + e^-$	2.71 v
$Mg \rightarrow Mg^{+2} + 2\ e^-$	2.37 v
$Al \rightarrow Al^{+3} + 3\ e^-$	1.66 v
$H_2 + 2\ OH^{-1} \rightarrow 2\ H_2O + 2\ e^-$	0.83 v
$Zn \rightarrow Zn^{+2} + 2\ e^-$	0.76 v
$Cr \rightarrow Cr^{+3} + 3\ e^-$	0.74 v
$Fe \rightarrow Fe^{+2} + 2\ e^-$	0.44 v
$Pb + SO_4^{-2} \rightarrow PbSO_4 + 2\ e^-$	0.31 v
$H_2 \rightarrow 2\ H^{+1} + 2\ e^-$	0.00 v
$Cu \rightarrow Cu^{+2} + 2\ e^-$	−0.34 v
$2\ I^{-1} \rightarrow I_2 + 2\ e^-$	−0.54 v
$Fe^{+2} \rightarrow Fe^{+3} + e^-$	−0.77 v
$Ag \rightarrow Ag^{+1} + e^-$	−0.80 v
$2\ Br^{-1} \rightarrow Br_2 + 2\ e^-$	−1.09 v
$2\ H_2O \rightarrow O_2 + 4\ H^{+1} + 4\ e^-$	−1.23 v
$2\ Cl^{-1} \rightarrow Cl_2 + 2\ e^-$	−1.36 v
$PbSO_4 + 2\ H_2O \rightarrow PbO_2 + 4\ H^{+1} + SO_4^{-2} + 2\ e^-$	−1.70 v
$2\ F^{-1} \rightarrow F_2 + 2\ e^-$	−2.87 v

$$Fe \rightarrow Fe^{+2} + 2\ e^{-1} \qquad E° = + 0.44\ v$$

$$Cu \rightarrow Cu^{+2} + 2\ e^{-1} \qquad E° = - 0.34\ v$$

This means that iron metal has a greater tendency to be oxidized than does copper metal, and thus iron will act as a reducing agent, reducing copper. The overall equation is presented below.

$$Fe \rightarrow Fe^{+2} + 2\ e^{-1}$$
$$\underline{Cu^{+2} + 2\ e^{-1} \rightarrow Cu}$$
$$Fe\ +\ Cu^{+2} \rightarrow Fe^{+2} + Cu$$

This is a balanced ionic equation. As with the ionic equations discussed earlier, in Chapter 8, the sulfate ion does not actually participate in the chemical reaction, but remains in solution unchanged. Evaporation of the aqueous solution would give ferrous sulfate ($FeSO_4$).

A close, direct association exists between redox reactions and electricity. In some instances, electricity can bring about redox reactions, as in electroplating and electrorefining. In other instances, redox reactions generate electricity, as in electrical cells and batteries. These processes will be discussed more fully in the next sections.

electroplating

In the previous discussion, spontaneous oxidation-reduction reactions involving metals were discussed. It is possible to make these reactions proceed in the reverse direction if electrical energy is supplied. This is accomplished by placing the two metals in a solution containing the salt of one and applying an electrical potential to each.

An entire industry, the electroplating industry, has developed on this principle. Chromium-plated automobile bumpers and silver-plated tableware are among the common items associated with this industry. Many other, less well-known items also are electroplated before being used in the household and in industry.

As a typical example, consider silver plating. A tank is set up as shown in Figure 10-1. The item to be silver-plated is connected to the *cathode* (negative electrode) of the power supply and a piece of pure silver is connected to the *anode* (positive electrode). A solution of a silver salt in water is placed in the tank. The solution is called an electrolyte because it conducts an electrical current. The choice of silver salt affects the properties of the resulting plating. Frequently

Figure 10-1. Apparatus for silver plating

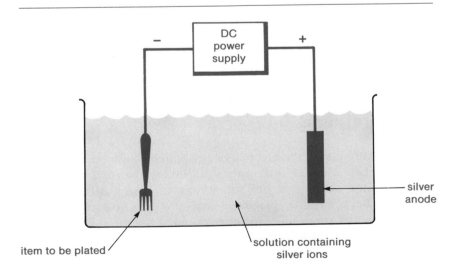

other inert compounds also are added, which can improve the quality of the plated materials. (The proper selection of additives is presently more an art than a science.)

When an electrical current is applied, a chemical reaction takes place at the electrodes. Reduction occurs at the cathode, where the item to be plated is located. Silver metal is deposited on the cathode, formed from silver ions in solution.

$$Ag^{+1} + e^{-1} \rightarrow Ag$$

At the same time, oxidation occurs at the anode, and the silver anode slowly dissolves.

$$Ag \rightarrow Ag^{+1} + e^{-1}$$

For every silver atom deposited on the object at the cathode, one silver atom is dissolved at the anode. The net overall reaction is the transfer of silver metal from the anode to the cathode.

electrorefining

The procedure just described has been applied to the purification of certain metals. Copper, for example, can be obtained reasonably pure by other techniques, but sometimes must be purified further. This is accomplished by placing a large piece of impure copper at the anode and a small piece of pure copper at the cathode. A copper salt solution is used to carry the current. As the reaction proceeds, the crude copper is dissolved, and pure copper is deposited on the

Cathodes of 99.98% pure copper are removed from an electrolytic refining tank. (*Courtesy Anaconda Company, New York*)

cathode. The relatively unreactive impurities collect as a sludge on the bottom of the electrolysis cell. Impurities such as silver, platinum, and gold are found in this sludge. The more reactive impurities dissolve in the electrolyte, but are not plated on the cathode because the electrical potential is adjusted so that they will stay in solution. The result is the deposition of a very pure block of copper at the cathode.

The more highly reactive elements do not occur freely in nature. Several of these elements have become important materials in industry, however. Among these are aluminum, magnesium, sodium, and chlorine. These elements can be produced in pure form by electrochemical techniques.

Aluminum is the most common metal in the earth's crust, but until relatively recently, it was also one of the most difficult to produce commercially. The main process used today is the Hall-Heroult method, which was developed around 1886. A diagram of the process is shown in Figure 10-2.

Both electrodes are made of carbon. Molten cryolite (Na_3AlF_6) is used as the electrolyte and is maintained at about 1000°C. A pure form of alumina (Al_2O_3) provides the source of aluminum for the process. Cryolite apparently causes the alumina to ionize, permitting successful electrolysis of aluminum.

(*Above left*) After electrorefining, pure aluminum is poured into molds to form ingots from which aluminum products then are manufactured. (*Courtesy Aluminum Company of America*)

(*Above right*) A huge rolling mill produces plates of aluminum. (*Courtesy Aluminum Company of America*)

Bauxite ore being excavated from an open-pit mine in Jamaica. (*Courtesy Aluminum Company of America*)

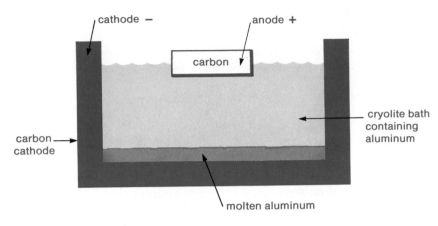

Figure 10-2   The Hall-Heroult process for producing aluminum

Aluminum is produced at the cathode according to the following equation.

$$Al^{+3} + 3\ e^{-1} \rightarrow Al$$

The oxygen that is liberated initially at the anode reacts with the carbon electrode so that the major product at the anode is carbon dioxide.

$$C + 2\ O^{-2} \rightarrow CO_2 + 4\ e^{-1}$$

The overall reaction is as follows.

$$2\ Al_2O_3 + 3\ C \rightarrow 4\ Al + 3\ CO_2$$

Enough heat is liberated by the electrolysis to maintain the temperature at the desired level without external heating. Molten aluminum collects on the bottom of the cell and is removed periodically.

The mineral source of alumina is bauxite. Bauxite is a hydrated form of alumina found naturally in several places around the world. It must be purified before electrolysis, to remove other metals like iron that also would be plated out with the aluminum if allowed to remain.

Sodium metal is produced by electrolysis of a mixture of molten sodium chloride and calcium chloride. This mixture is used because it has a lower melting point than pure sodium chloride. Reduction of the sodium ion takes place at one carbon electrode to produce sodium metal.

$$Na^{+1} + e^{-1} \rightarrow Na$$

Since molten sodium is lighter than sodium chloride, it floats to the surface of the electrolyte bath where it is separated and allowed to solidify.

Chlorine gas is produced by oxidation at the carbon electrode.

$$2\ Cl^{-1} \rightarrow Cl_2 + 2\ e^{-1}$$

This process is not the preferred source of commercial quantities of chlorine gas, however. Chlorine gas is prepared more conveniently by electrolysis of an aqueous solution of sodium chloride.

Chlorine is used extensively in industry for everything from water purification to manufacture of shower curtains. Most chlorine is produced by the electrolysis of aqueous sodium chloride solution.

In this electrolysis, the anode reaction proceeds as in the electrolysis of molten sodium chloride with the production of chlorine gas.

$$2\ Cl^{-1} \rightarrow Cl_2 + 2\ e^{-1}$$

Sodium metal is not produced at the cathode, however. Data in Table 10-1 show that water is reduced more easily than is the sodium ion. As a consequence, the cathode reaction is as follows.

$$2\ H_2O + 2\ e^{-1} \rightarrow H_2 + 2\ OH^{-1}$$

Thus hydrogen gas forms at the cathode rather than sodium metal, and sodium hydroxide is left behind in the cell. The overall chemical reaction is given below.

$$2\ NaCl + 2\ H_2O \rightarrow Cl_2 + H_2 + 2\ NaOH$$

The reaction vessel in which this electrolysis is conducted is divided into two compartments by an asbestos diaphragm, causing the sodium hydroxide to concentrate in the cathode compartment. The solution in this compartment is drawn off continuously for concentration and removal of the sodium hydroxide from the unreacted salt, which then can be recycled into the electrolysis process. Figure 10-3 illustrates this procedure.

An alternate process for chlorine production operates essentially the same way, but employs a mercury cathode. For economic reasons, this has been a popular commercial process. In recent years, several plants have been shut down in the Great Lakes area because mercury

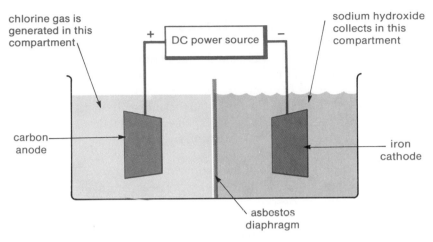

chlorine gas is generated in this compartment

sodium hydroxide collects in this compartment

+ DC power source −

carbon anode

iron cathode

asbestos diaphragm

Figure 10-3. The production of chlorine in a diaphragm cell

from the electrolysis units was escaping into the lakes and has been found in fish. Mercury, like most metals, is highly toxic. The economic advantage of this process now is being reevaluated because of the previously unrecognized need to prevent the escape of mercury into the country's natural waterways. The problem of mercury contamination is discussed further in Chapter 18.

Electrochemical diaphragm cells for the production of chlorine gas from aqueous sodium chloride. (*Courtesy Hooker Chemicals and Plastics Corporation, Niagara Falls, N.Y.*)

voltaic cells and batteries

In an earlier section, it was stated that certain metals will spontaneously displace other metal ions from solution. There also has been established an order of reactivity that can predict which metals will displace which other metal ions. Furthermore, it has been stated that it is possible to make these displacement reactions proceed in the reverse (nonspontaneous) direction by applying electrical energy to the system.

This section will show how oxidation-reduction reactions can be used to produce electrical energy. To accomplish this, three voltaic cells will be described; these cells are, or have been, commonly used. A voltaic cell is an apparatus that is designed to produce electricity from a chemical reaction. It is named for the discoverer, Alessandro Volta.

In the early days of the telegraph, modern electrical power plants were unknown. Even the common dry cell and automobile battery had yet to be developed. At that time, each telegraph office was equipped with a gravity cell, which the telegraph operator assembled and maintained himself. This cell provided electricity for the functioning of the telegraph apparatus.

The gravity cell consisted of a glass jar containing a copper electrode on the bottom and a zinc electrode near the top. The copper electrode was immersed in a saturated solution of copper sulfate. A dilute solution of zinc sulfate, in which the zinc electrode was immersed, was poured on top of the copper sulfate solution carefully, so that the two solutions did not mix. A gravity cell is shown in Figure 10-4.

If the zinc electrode had been placed directly into the copper sulfate solution, the electrode would have dissolved and copper metal would have separated out, according to the following equation.

$$Zn \rightarrow Zn^{+2} + 2\ e^{-1}$$
$$\underline{Cu^{+2} + 2\ e^{-1} \rightarrow Cu}$$
$$Zn\ +\ Cu^{+2} \rightarrow Zn^{+2} + Cu$$

No electrical energy would be available for use because of the direct contact between the reactants. This reaction cannot occur directly in the gravity cell, however, because the zinc electrode is separated physically from the copper sulfate solution.

Nevertheless, a redox reaction will occur if an electrical connection is made between the two electrodes. When the zinc metal dissolves, zinc atoms are converted to zinc ions, which go into

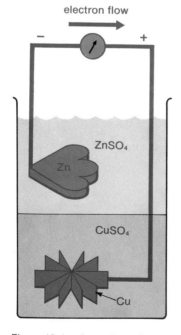

electron flow

ZnSO₄

Zn

CuSO₄

Cu

Figure 10-4.   A gravity cell

solution. Each zinc atom leaves its two electrons behind on the zinc electrode.

$$Zn \rightarrow Zn^{+2} + 2\ e^{-1}$$

These electrons are then conducted through the electrical connection to the copper electrode, where they are transferred to copper ions from the copper sulfate solution.

$$Cu^{+2} + 2\ e^{-1} \rightarrow Cu$$

The copper atoms are deposited on the copper electrode. Thus the zinc electrode slowly dissolves while the copper electrode slowly increases in size. The net chemical reaction is the same as if the zinc had been added directly to the copper sulfate solution.

$$Zn + Cu^{+2} \rightarrow Zn^{+2} + Cu$$

No reaction occurs unless the external electrical circuit is closed. The flow of electrons within the circuit constitutes the electrical current that the cell provides. The cell's voltage relates to the oxidation potentials of the half reactions and the concentration of the solutions. The current-carrying capacity of the cell relates to the surface area of the electrodes. When the zinc electrode dissolves nearly completely or the copper sulfate concentration diminishes significantly, the cell goes "dead."

The gravity cell had two major disadvantages. It was messy to work with, and it was not portable. A more recent development is the dry cell, which is used in flashlights and portable radios. A cutaway drawing of the essential features of a dry cell is given in Figure 10-5.

The dry cell is not actually dry. It contains a paste made of water, ammonium chloride, and manganese dioxide. The top is sealed, however, to prevent these materials from leaking. The case is made of zinc metal, which also functions as the negative electrode. The carbon rod in the center functions as the positive electrode.

Oxidation of the zinc case produces electrons in the same manner as in the gravity cell.

$$Zn \rightarrow Zn^{+2} + 2\ e^{-1}$$

The reduction reaction at the carbon electrode involves both water and manganese dioxide.

$$2\ MnO_2 + 2\ H_2O + 2\ e^{-1} \rightarrow 2\ MnO(OH) + 2\ OH^{-1}$$

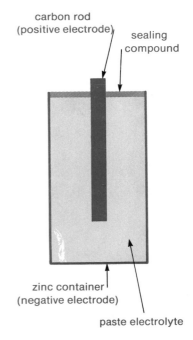

carbon rod (positive electrode)

sealing compound

zinc container (negative electrode)

paste electrolyte

Figure 10-5.   A dry cell

The ammonium chloride that is present serves two purposes. First, it prevents the solution from becoming excessively alkaline by reacting with the hydroxide ion.

$$NH_4^{+1} + OH^{-1} \rightarrow NH_3 + H_2O$$

Second, the ammonia produced in this last reaction combines with the zinc ions to prevent a buildup of zinc ions in the mixture.

$$Zn^{+2} + 2\ NH_3 + 2\ Cl^{-1} \rightarrow Zn(NH_3)_2Cl_2$$

The net overall reaction is the following equation.

$$Zn + 2\ MnO_2 + NH_4Cl \rightarrow 2\ MnO(OH) + Zn(NH_3)_2Cl_2$$

The lead storage battery used in automobiles and trucks consists of a group or battery of lead storage cells connected in series. Each cell produces about two volts of electricity. Thus, a 6-volt battery contains three cells and a 12-volt battery contains six cells.

Each cell consists of alternating lead and lead dioxide plates separated by wooden or plastic spacers. The lead plates are all connected in parallel and constitute the negative electrode. The lead dioxide plates also are connected in parallel and compose the positive electrode. The current-producing capacity of the cell relates to the number and size of the plates. An aqueous solution of sulfuric acid serves as the electrolyte. A diagram of a portion of a lead storage cell appears in Figure 10-6.

Figure 10-6.   A portion of a lead storage cell

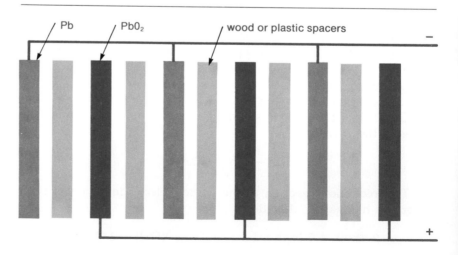

When the cell is being discharged (supplying electricity), oxidation occurs at the lead electrode according to the following half reaction.

$$Pb + SO_4^{-2} \rightarrow PbSO_4 + 2\ e^{-1}$$

At the same time, reduction occurs at the lead dioxide electrode.

$$PbO_2 + 4\ H^{+1} + SO_4^{-2} + 2\ e^{-1} \rightarrow PbSO_4 + 2\ H_2O$$

Thus the sulfuric acid is consumed, and both electrodes are converted to insoluble lead sulfate. The overall chemical reaction is as follows.

$$Pb + PbO_2 + 2\ H_2SO_4 \rightarrow 2\ PbSO_4 + 2\ H_2O$$

The lead storage battery has the advantage that it can be recharged readily, by applying an electrical potential to the battery strong enough to reverse the chemical reaction described above.

$$2\ PbSO_4 + 2\ H_2O \xrightarrow{DC} Pb + PbO_2 + 2\ H_2SO_4$$

Recharging the battery is the function of a car generator or battery charger.

study questions

1 Name several oxidation-reduction reactions that you are likely to see outside a chemical laboratory (for example, in the home or outdoors).

2 If a substance becomes oxidized in a chemical reaction, must there likewise be a substance that becomes reduced? Why?

3 Determine the oxidation numbers of the italicized elements in each of the following species.
   a. K d. N_2O_4 f. Zn^{+2}
   b. SO_2 e. $KHCO_3$ g. MnO_4^{-1}
   c. HCO_2H

4 Indicate which species is oxidized and which is reduced in each of the following reactions.
   a. $C_3H_8 + 5\ O_2 \rightarrow 3\ CO_2 + 4\ H_2O$
   b. $2\ Na + 2\ H_2O \rightarrow 2\ NaOH + H_2$
   c. $2\ FeCl_2 + 2\ HCl \rightarrow 2\ FeCl_3 + H_2$

5 Identify the oxidizing agents and the reducing agents in the reactions in Question 4.

6 For what purpose are equations written for half reactions?

7 Potassium has a higher oxidation potential (2.93 volts) than sodium (2.71 volts). Is potassium therefore more easily oxidized than sodium? Which element would be a better reducing agent?

8 Using information provided in Table 10-1, predict which of the following reactions would be spontaneous.
a. $Mg + Zn^{+2} \rightarrow Mg^{+2} + Zn$
b. $Cu + 2\ Ag^{+1} \rightarrow Cu^{+2} + 2\ Ag$
c. $Fe + Zn^{+2} \rightarrow Fe^{+2} + Zn$
d. $2\ I^{-1} + Cl_2 \rightarrow I_2 + 2\ Cl^{-1}$
e. $2\ Cl^{-1} + Br_2 \rightarrow Cl_2 + 2\ Br^{-1}$

9 If a reaction is spontaneous in the forward direction, is it also spontaneous in the reverse direction?

10 Sketch a labeled diagram of the apparatus that would be used to chrome-plate a cigarette lighter.

11 Calcium may be obtained commercially from $CaCl_2$ by an electrolytic process. Describe or illustrate how this might be be done, and indicate what other product would be formed.

12 Why does an automobile battery run down?

13 How is an automobile battery recharged? What happens in a battery when it is recharged?

special problems

1 a. Write a half reaction for the oxidation of magnesium.
b. Write a half reaction for the oxidation of copper to Cu^{+2}.
c. Magnesium has a higher oxidation potential than copper. In which case would you expect to see a chemical reaction: by placing magnesium metal in copper sulfate solution, or by placing copper metal in magnesium sulfate solution?
d. Write a balanced chemical equation for the reaction you selected in (c).

2 The chemical reaction that takes place during the discharge of an automobile battery is shown below. If the lead plate weighs 150 grams, how much must the lead oxide (PbO_2) plate weigh for complete reaction to be possible? Show the procedure used to arrive at your answer. (Review Chapter 9 if necessary.)

$$2\ H_2SO_4 + Pb + PbO_2 \rightarrow 2\ PbSO_4 + H_2O$$

3 In Problem 2, which plate is oxidized and which is reduced?

  Pb_____   PbO$_2$_____

4 Why is it no longer possible to recharge an automobile battery after several years of its operation?

5 Identify the oxidized and the reduced species in the following reactions.

  a. $Mn^{+2} + BiO_3^{-1} \rightarrow MnO_4^{-1} + Bi^{+3}$

  b. $I_2 + ClO_4^{-1} + H_2O \rightarrow IO_3^{-1} + Cl^{-1} + H^{+1}$

  c. $UO^{+2} + Cr_2O_7^{-2} \rightarrow UO_2^{+2} + Cr^{+3}$

chapter **11**

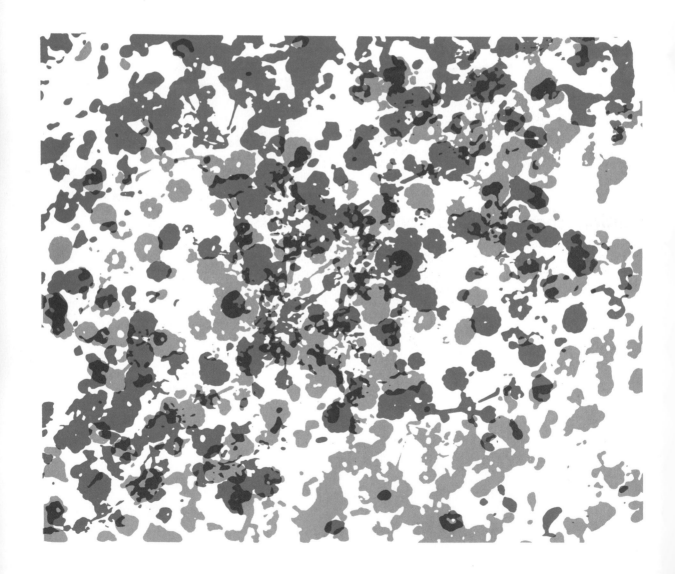

acids and bases

Everyone is familiar with substances such as vinegar, household ammonia, carbonated soft drinks, and bicarbonate of soda. The sour taste of citrus juices, the bitterness and slipperiness of soap solutions, and the corrosion of automobile batteries are equally familiar observations. The substances responsible for the properties of these household products are acids and bases. The characteristic properties of acids and bases are discussed in this chapter, and since acid-base chemistry takes place in solution, the chapter begins with a brief discussion of solutions.

solution chemistry and concentration units

Solutions are mixtures in which one substance, the solute, is dissolved in another, more abundant component, the solvent. Solutions are physically homogeneous and can be formed from substances of all three states of matter. Those in which liquids serve as solvents are the most common, however. Such materials as gasoline, milk, air, and alloys are examples of solutions.

Concentrations of solutions usually are expressed in terms of the number of moles of solute dissolved in one liter of solution. The concentration of a solution in these units is termed its molarity. Thus, a solution containing 0.5 mole of NaCl dissolved in one liter of solution would be called a 0.5 molar, or a 0.5 M, NaCl solution. Similarly, in a 6-M solution of sulfuric acid, six moles of sulfuric acid are dissolved in one liter of solution. Molar concentrations also can be expressed for ionic species, as for example a solution that is 0.25 M in nitrate ion (NO_3^{-1}).

Most ionic substances, when dissolved in water, dissociate into their respective ions. For example, sodium chloride dissociates completely in water to form sodium and chloride ions.

solution
solute
solvent
molarity
Arrhenius acid
Arrhenius base
Brönsted-Lowry acid
Brönsted-Lowry base
Lewis acid
Lewis base
neutralization reaction
salt
strong acid
moderate acid
weak acid
strong base
weak base
neutral
ion product of water
acid-base indicator
pH meter
buffer
hydronium ion

$$NaCl \xrightarrow{\text{H}_2\text{O}} Na^{+1} + Cl^{-1}$$

Some covalently bonded materials also tend to dissociate in water. Hydrogen chloride and sulfuric acid are common examples of this phenomenon.

$$HCl \xrightarrow{\text{H}_2\text{O}} H^{+1} + Cl^{-1}$$

$$H_2SO_4 \xrightarrow{\text{H}_2\text{O}} H^{+1} + HSO_4^{-1} \xrightarrow{\text{H}_2\text{O}} H^{+1} + SO_4^{-2}$$

The sulfate ion, like most complex ions, normally does not dissociate further in water. In most liquid solutions, the solvent can be removed by evaporation and the solute recovered unchanged.

Ionization of substances in an aqueous solution is responsible for the electrical conductivity of ionic materials in aqueous solutions. Without this phenomenon, electroplating, batteries, and other useful processes and devices would not be possible. Ionization also is responsible for the danger involved with a person's turning on electric lights while he is standing in a bathtub filled with water.

Occasionally, concentrations of solutions are expressed as percent by weight or volume. The former generally is used when the substances being dissolved (solutes) are solid; the latter usually is preferred for liquid solutes. Thus a 5% aqueous solution of sodium hydrogen carbonate contains 5 grams of sodium hydrogen carbonate in 95 grams of water, and a 45% aqueous ethyl alcohol solution contains 45 milliliters of ethyl alcohol in 55 milliliters of water.

Other ways of expressing solution concentration have been devised, but are not used so commonly as those just described.

acid-base theories

Three major theories of acids and bases have been proposed and used throughout the history of chemistry, and each successive theory has been broader in scope than the previous one. Based on the most general theory, it has been suggested that any chemical reaction that is not an oxidation-reduction reaction is an acid-base reaction.

The earliest and most fundamental concept of acids and bases was developed by a Swedish chemist, Svante Arrhenius. In 1887, Arrhenius defined an acid as a substance that ionizes in water to produce hydrogen ions. A base was defined by the Arrhenius concept as a substance that ionizes in water to produce hydroxide ions.

Examples of Arrhenius acids and bases are given below.

Acids: $HCl \xrightarrow{H_2O} H^{+1} + Cl^{-1}$

      Hydrochloric acid
      (muriatic acid)

$H_2SO_4 \xrightarrow{H_2O} 2\,H^{+1} + SO_4{}^{-2}$

    Sulfuric acid
    (battery acid)

Bases: $NaOH \xrightarrow{H_2O} Na^{+1} + OH^{-1}$

     Sodium hydroxide
     (soda lye)

$Mg(OH)_2 \xrightarrow{H_2O} Mg^{+2} + 2\,OH^{-1}$

  Magnesium hydroxide
  (milk of magnesia)

Although the Arrhenius theory of acids and bases is the least comprehensive of the three that have been proposed, it is nevertheless still widely used and is the basis for several concepts related to the properties of acidity and basicity. These concepts will be discussed later in the chapter.

In 1923, J. N. Brönsted, a Danish scientist, and T. M. Lowry, a British scientist, independently proposed a concept of acids and bases that expanded the ideas of Arrhenius. These two scientists suggested that an acid is any substance donating hydrogen ions (protons) in a chemical reaction and that a base is any substance accepting hydrogen ions (protons) in a reaction. According to the Brönsted-Lowry theory, then, HCl and H_2SO_4, which are typical Arrhenius acids, also can be called Brönsted-Lowry acids. In addition, such species as the ammonium ion ($NH_4{}^{+1}$) and water would be Brönsted-Lowry acids in reactions like the following.

$$NH_4{}^{+1} + OH^{-1} \rightarrow NH_3 + HOH$$

$$H_2O + CH_3O^{-1} \rightarrow OH^{-1} + CH_3OH$$

The Brönsted-Lowry concept classifies species such as OH^{-1} and CH_3O^{-1} in the reactions above as bases, since they serve as proton acceptors. Further discussion of the Brönsted-Lowry concept will appear later in the chapter.

The broadest theory of acids and bases was proposed by G. N. Lewis, an American chemist also recognized for developing the concept of covalent bonding. Lewis defined an acid as a species that accepts an electron pair in a chemical reaction and a base as a species that donates an electron pair in a reaction. An example of a Lewis acid-base reaction is given below.

$$BF_3 + :NH_3 \rightarrow F_3B-NH_3$$

Lewis acid   Lewis base

$$:\overset{..}{\underset{..}{F}}: \qquad H \qquad :\overset{..}{\underset{..}{F}}:H$$
$$:\overset{..}{\underset{..}{F}}:B \;+\; :N:H \;\rightarrow\; :\overset{..}{\underset{..}{F}}:B:N:H$$
$$:\overset{..}{\underset{..}{F}}: \qquad H \qquad :\overset{..}{\underset{..}{F}}:H\cdot$$

The boron atom in BF_3 has a vacant electron orbital that can accept a pair of unbonded electrons on the nitrogen atom in NH_3. A new covalent bond forms in the product that serves to bond the two molecules together as an acid-base reaction product.

The Lewis concept of acids and bases will not be discussed further in this book. The following sections describing other aspects of acid and base chemistry are based on the Arrhenius and Brönsted-Lowry theories.

neutralization reactions

A common remedy for indigestion is bicarbonate of soda or a similar commercial preparation such as Alka-Seltzer. Indigestion often results from excess acid in the stomach due to oversecretion of acidic gastric juices. Substances such as bicarbonate of soda and Alka-Seltzer are bases; they relieve the discomfort of indigestion by reducing the acidity of the stomach content through a chemical reaction. Such acid-base reactions are called neutralization reactions and produce compounds called salts. Examples of neutralization reactions between Arrhenius acids and bases are given below.

$$HCl + NaOH \rightarrow NaCl + H_2O$$

$$H_2SO_4 + Mg(OH)_2 \rightarrow MgSO_4 + 2\,H_2O$$
(Milk of magnesia)

$$3\,HNO_3 + Al(OH)_3 \rightarrow Al(NO_3)_3 + 3\,H_2O$$
(Gelusil)

It can be readily noted that these reactions have a common product, water. This is characteristic of all reactions between Arrhenius acids and bases. Since all these reactions are conducted in

Neutralization of vinegar (an acid) with bicarbonate of soda (a base). (*Used with permission of Earlham College Press, copyright owner, and McGraw-Hill Book Company, publisher.*)

aqueous media and since most resulting salts remain in solution as ionic species, the only true product is water. Therefore, it often is stated that the only actual reaction occurring between an Arrhenius acid and an Arrhenius base is the following.

$$H^{+1} + OH^{-1} \rightarrow H_2O$$

Such a reaction would account for the lessening of the acidic and basic characteristics of the reactants. Evaporation of the water from neutralization-reaction mixtures would yield residues of the various salts.

acid and base strengths

The acidic strengths or properties of compounds are not all equal, nor are their basic strengths or properties. Likewise, the acidity or basicity of a solution can vary, depending on several factors. It is therefore necessary to develop some concepts or criteria for establishing relative acidities and basicities of matter.

The acidic nature of an Arrhenius acid is based on the degree of ionization of the substance in water. For example, hydrochloric acid (or hydrogen chloride) is ionized completely in water and so is referred to as a strong acid. Other strong acids include nitric acid (HNO_3) and sulfuric acid (H_2SO_4). Phosphoric acid (H_3PO_4) is less ionized in water and is classified as a moderate acid. Carbonic acid (H_2CO_3), the acid in carbonated drinks, and acetic acid ($HC_2H_3O_2$), the acid in vinegar, are only slightly ionized in water and are called weak acids. These acids exist more predominantly as molecules rather than as ions in water.

The same general criterion applies to strengths of bases. Strong bases, those that ionize completely in water, include sodium hydroxide (NaOH) and potassium hydroxide (KOH), but ammonium hydroxide (NH_4OH), which partially ionizes, is termed a weak base.

When the terms *acidity* and *basicity* are applied to aqueous solutions of substances, the criterion determining these properties is the relative concentration of hydrogen ions to hydroxide ions. All aqueous solutions contain both these ions in varying, relative amounts. Whichever ion is in greater concentration will determine the overall acidic or basic nature of the solution. The greater the difference between the hydrogen ion and hydroxide ion concentrations, the more acidic or basic the solution (depending on which ion predominates). In pure water the concentrations of both ions are equal; therefore, pure water is neither acidic nor basic, but is neutral. It has been observed that in all aqueous solutions the mathematical product of the molar concentration of the hydrogen ions multiplied by the molar concentration of the hydroxide ions is a constant, 1×10^{-14}. This constant is known as the ion product of water, K_w.

$$K_w = [H^{+1} \text{ conc.}] \, [OH^{-1} \text{ conc.}] = 1 \times 10^{-14}$$

To express the acidity or basicity of a solution in more absolute terms, a numerical scale has been developed. This scale is the pH scale, and it goes from 0 to 14. The numbers in the scale relate to the hydrogen ion concentration in an aqueous solution. The divisions in the pH scale are shown below.

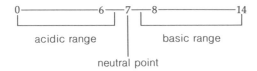

The more acidic a solution is, the lower is its pH, and the more basic it is, the higher is its pH. When a solution is neutral, its pH is 7. The pH values of some familiar solutions are given in Table 11-1.

acid-base indicators

The acidic or basic character of a solution can be determined by the use of substances called acid-base indicators. These substances, which themselves are weak acids or bases, assume one color below a

Table 11-1.   Average pH values of some solutions

Stomach content	2
Lemon juice	2.3
Vinegar	2.9
Soda pop	3
Milk	6.5
Human blood plasma	7.4
Milk of magnesia	10
Household ammonia	11
Washing soda	12

Swimming pools are usually "chlorinated" to kill harmful organisms. The material that is actually added is not normally chlorine gas, but sodium hypochlorite (NaOCl). The chlorine in sodium hypochlorite destroys organisms, and the rest of the molecule is converted to sodium hydroxide.

$$2 \ NaOCl + H_2O \rightarrow 2 \ Cl + 2 \ NaOH$$

Over time, the sodium hydroxide concentration in the water increases to the point that the pH of the solution becomes too high for safe swimming. Muriatic acid (industrial hydrochloric acid) then is added in small quantities to lower the pH to near neutral (pH of 7).

$$HCl + NaOH \rightarrow H_2O + NaCl$$

certain pH range and another color above that pH range. Each inditor therefore makes the transition in color over a specific pH range.

One of the most common and best known acid-base indicators is litmus, which is red in acidic solutions and blue in basic solutions. Another frequently used indicator is phenolphthalein, which is colorless in acidic solutions and red in basic solutions. Table 11-2 lists some acid-base indicators and their colors at different pH ranges.

Table 11-2.   Some common acid-base indicators

Indicator	Acidic Color	Basic Color	pH Range for Color Change
Methyl orange	red	orange	3–4.4
Congo red	blue	red	3–5
Litmus	red	blue	5–8
Bromthymol blue	yellow	blue	6–7.5
Phenolphthalein	colorless	red	8.3–10
Alizarin yellow	yellow	lilac	10–12

Above: Portable battery-operated pH meter used to measure quickly and accurately a solution's pH. *Below*: Similar laboratory-model instrument. (*Courtesy Beckman Instruments, Inc.*)

A number of acid-base indicators are substances that are obtained from natural materials. One of the most commonly used indicators, litmus, is a product of lichens which grown on rocks and tree trunks. Red cabbage contains a pigment that can also be used as an acid-base indicator. A solution of this pigment can be made by soaking chopped-up red cabbage leaves in hot water for about an hour. The purple solution that is obtained will turn reddish when treated with an acid such as vinegar. When treated with a base such as household ammonia, it will turn bluish and then green.

Several "universal" indicator solutions have been formulated that permit determination of the pH of any aqueous solution over nearly the entire pH scale. Such indicator solutions are actually mixtures of several indicators, each of which changes color in a different pH range. Specially prepared indicator test papers and strips also are available for determining pH values of solutions. Indicators are useful in many phases of laboratory chemical analyses and in such procedures as water-quality control and urinalysis.

The use of acid-base indicators has certain limitations. Measuring a solution's pH may be made difficult by the color of the solution itself, which could mask the indicator color. Persons who are color-blind also have difficulty interpreting subtle color shades and changes. For the most accurate measurement of pH, instruments called pH meters are used. These instruments determine the pH of a solution by measurement of the voltage between two electrodes immersed in the solution. The electrical potential between the electrodes is brought about by the hydrogen ions in solution.

buffers

When an aqueous solution of a weak acid is mixed with its salt, the resulting solution is called a buffer. A buffer solution maintains an essentially constant pH (or hydrogen ion concentration), even when another acid or base is added to it. For example, a solution containing equal quantities of acetic acid and sodium acetate will have a pH of 4.74. Addition of moderate amounts of hydrochloric acid or sodium hydroxide will affect the pH of this solution only slightly because components within the solution can react with the added acid or base to neutralize it. The acetic acid in the buffer will react with any base that is added to neutralize it, and the acetate ion from the sodium acetate will react with any acid that is added to neutralize it. The following equations illustrate the reactions that occur.

$$HC_2H_3O_2 \; + \; OH^{-1} \; \rightarrow \; H_2O \; + \; C_2H_3O_2^{-1}$$
(from
added base)

$$C_2H_3O_2^{-1} \; + \; H^{+1} \; \rightarrow \; HC_2H_3O_2$$
(from   (from   (weak acid)
sodium acetate)   added acid)

Buffers commonly are used in chemistry to control the pH of solutions. All of these buffers work essentially in the same manner as the acetic acid/sodium acetate buffer described earlier. Sodium phosphate in household detergents functions as a buffer to maintain the pH of laundry water at an optimum value for washing clothes. Nature employs buffers extensively in living cells and higher organisms. Blood, for example, is buffered at a pH of about 7.4, which is nearly neutral. The absorption of carbon dioxide and oxygen by the blood is less efficient at either higher or lower pH values.

the brönsted-lowry concept

As defined earlier, a Brönsted-Lowry acid is a *proton donor* in a chemical reaction, and a Brönsted-Lowry base is a *proton acceptor*. Consequently, in the reaction,

$$HCl + NH_3 \rightarrow NH_4^{+1} + Cl^{-1}$$

HCl would be called an *acid* and NH_3 a *base*.

The Brönsted-Lowry concept is especially useful in attempting to describe what actually happens when an acid ionizes in water. When hydrogen chloride molecules dissolve in water, for example, they do not merely ionize to form hydrogen and chloride ions as has been implied earlier. Instead, an acid-base interaction occurs with the water molecules, as illustrated below.

$$HCl + H_2O \rightarrow Cl^{-1} + H_3O^{+1}$$

The **hydronium ion** (H_3O^{+1}) results from the acceptance by a water molecule of a proton from the HCl molecule. Water therefore is behaving as a Brönsted-Lowry base. Likewise, all aqueous solutions of acids contain hydronium ions rather than free protons.

generality of acid-base phenomena

As shown in this chapter, acids and bases may be defined very generally or very specifically. Regardless of which definition is used, acids and bases occur throughout the natural and man-made world. In some of the following chapters, acids and bases will be found responsible for such diverse activities as cleansing action in detergents, leavening action in baking, and deterioration of marble statues and buildings in smoggy cities.

study questions

1 Name five liquid solutions, three gaseous solutions, and three solid solutions.
2 What are some ways in which concentrations of solutions may be expressed?
3 Pure distilled water does not conduct an electrical current, but salt solution does. Why is this so?
4 Label each of the following compounds as either an acid or a base.
  a. NaOH   d. $Al(OH)_3$
  b. HNO_3 e. H_2SO_4
  c. H_3PO_4 f. $Ca(OH)_2$
5 Why is a reaction between an acid and a base called a neutralization reaction?
6 It has been stated that the only reaction actually occurring between an Arrhenius acid and an Arrhenius base is the formation of water from hydrogen and hydroxide ions. Why is this statement correct?
7 What distinguishes a strong acid from a weak acid?
8 Would solutions with the following pH values be acidic, basic, or neutral?
  a. 5   b. 12   c. 7   d. 9   e. 3
9 Which of the solutions in Question 8 is the most acidic? Which is the most basic?
10 If some KOH were added to an HNO_3 solution, would the pH of the solution increase, decrease, or remain the same? Why? Would the same result be obtained if KCl were added instead of KOH? Why?

11 Periodically, muriatic acid (HCl) must be added to swimming pool water to maintain the desired pH. Why is the addition necessary?

12 According to information given in Table 11-2, what colors would bromthymol blue, phenolphthalein, and methyl orange have in solutions with the following pH's?

a. 5   b. 12   c. 2   d. 8

13 Living cells contain a phosphate buffer. What is the importance of the buffer to the proper functioning of the cell?

14 Name at least one commercial product that contains a buffer.

special problems

1 A 1.5-molar solution of H_3PO_4 contains how many moles of H_3PO_4 in 1 liter of solution?

2 How many grams of H_3PO_4 are contained in one liter of the solution in Problem 1?

3 How many grams of CH_3OH are contained in 200 milliliters of a 3-molar solution?

4 Fifteen grams of NaOH are dissolved in enough water to make 500 milliliters of solution. What is the molarity of this solution?

5 How would you prepare a 30% aqueous sugar solution?

6 Identify the Brönsted-Lowry acid and base in each of the following reactions.

a. $HC_2H_3O_2 + OH^{-1} \rightarrow C_2H_3O_2^{-1} + H_2O$

b. $SO_4^{-2} + H_3O^+ \rightarrow HSO_4^{-1} + H_2O$

c. $NH_4^{+1} + NH_2^{-1} \rightarrow 2\ NH_3$

d. $H_2O + NH_3 \rightarrow NH_4^{+1} + OH^{-1}$

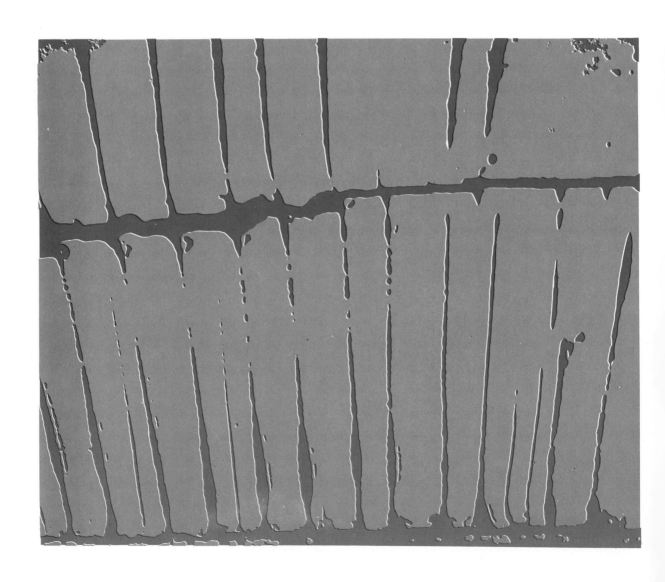

carbon—a unique element

In its original interpretation, organic chemistry was the study of the compounds resulting from life processes. It was believed originally that only living systems could produce organic molecules. This concept was shown to be incorrect, and many synthetic organic compounds now have been made. Today, organic chemistry is identified as the chemistry of carbon compounds. Study of the chemistry of life processes is now the realm of biochemistry, which is discussed in the next chapter.

carbon

Carbon is probably the most important of all known elements. It has four electrons more than the next smaller inert gas, helium, and four electrons less than the next larger inert gas, neon. Carbon tends to form covalent bonds rather than to gain or lose electrons to form ionic bonds. Not only does carbon form covalent bonds with other elements such as hydrogen and oxygen, but also it forms extensive chains through covalent bonds with itself. This property is almost unknown among other elements. Because of this property, nearly two million compounds of carbon are known; they vary in physical properties from gases to crystalline solids and from highly reactive compounds to practically inert ones.

In most compounds, the angles between the four bonds on a carbon atom are equal. The configuration that permits this is the tetrahedral configuration. Thus carbon is visualized as located at the center of a tetrahedron. The four atoms or groups bonded directly to carbon then are located at the tetrahedron's four corners. Some three-dimensional representations of tetrahedral carbon are given in Figure 12-1.

As shown in Chapter 6, the electron configuration of carbon is

organic chemistry
biochemistry
orbital hybridization
isomer
optical isomer
racemic mixture
alkane
saturated hydrocarbon
paraffin
normal alkane
straight-chain paraffin
structural isomer
geometric isomerism
alkene
olefin
petroleum
fractional distillation
catalytic cracking
catalyst
functional group
carbonyl group

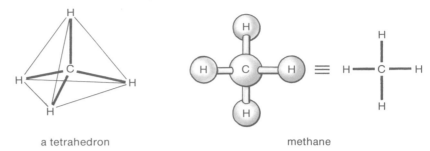

a tetrahedron   methane

Figure 12-1.   Some representations of a tetrahedral carbon

Model of the famous Hope Diamond in the American Museum of Natural History. Diamonds, the hardest natural substances known, are a form of pure carbon in which atoms occur in a rigid symmetrical lattice that extends in three dimensions. (*Courtesy American Museum of Natural History*)

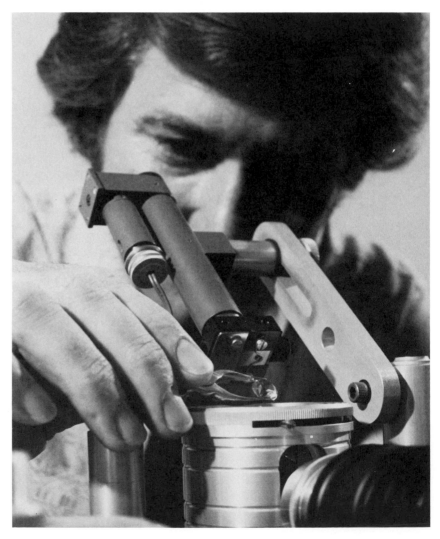

Jeweler examines a gem. Diamonds are one of the world's most popular jewels. (*UPI Photo*)

$1s^2 2s^2 2p^2$, but this is inconsistent with the tetrahedral geometry observed for carbon compounds. This specific geometry could be predicted, however, if the $2s$ and three $2p$ orbitals of carbon reorganize to form four hybrid orbitals of equal energy. The four new orbitals that result are known as sp^3 orbitals because they represent a composite of one s and three p orbitals. The process by which these combined or hybrid orbitals are produced is known as orbital hybridization. The four sp^3 hybridized orbitals assume a tetrahedral orientation, with the carbon nucleus at the center of the tetrahedron, as seen at the right.

It is difficult to draw the tetrahedral structure on paper, especially for compounds that have large and complicated structures. Consequently, organic structures usually are drawn as planar projections of the three-dimensional tetrahedral structures. Thus

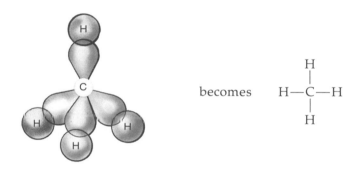

becomes

$$\begin{array}{c} H \\ | \\ H-C-H \\ | \\ H \end{array}$$

This representation may be somewhat difficult to interpret at first until it becomes thoroughly familiar. On paper, for example, there appears to be a difference between the following structures.

$$\begin{array}{c} H \\ | \\ H-C-Cl \\ | \\ Cl \end{array} \quad \text{and} \quad \begin{array}{c} Cl \\ | \\ H-C-H \\ | \\ Cl \end{array}$$

These structures, however, both represent the same compound, dichloromethane, which has only one actual structure. The identity of these structures can be shown by considering and comparing the three-dimensional representations of the two molecules, as shown at the right.

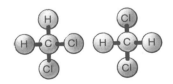

Both these structures can be rotated so that they superimpose. Thus they are identical.

optical isomers

The situation becomes just a little more complicated when there are four different groups on the carbon atom. In this case it is possible to consider two molecules that are mirror images of one another.

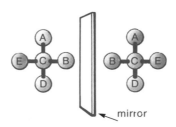

These two molecules are not superimposable. Compounds that have the same number and kinds of atoms, but which are arranged differently, are called isomers. Isomers are found extensively among organic compounds. When the only difference between isomers is that one isomer represents the mirror image of the other, the compounds are called optical isomers. Another kind of isomerism, to be discussed later, is structural isomerism. Most isomers differ significantly in physical properties. Optical isomers, however, are identical in all properties except for their behavior toward plane-polarized light (light vibrating in only one plane). One optical isomer will rotate plane-polarized light to the right; the other will rotate the light to the left. An equimolar mixture of the two optical isomers is called a racemic mixture and exhibits no net rotation of light.

Optical isomers have been used in studies of the way organic molecules react. They also occur extensively in living systems. Amino acids, from which proteins are made, are optically active (exist as optical isomers), as are sugars. These compounds are discussed further in Chapter 13.

Mirror imagery is demonstrated by a pair of shoes. Optical isomers show a similar pattern of mirror imagery. (*Photograph by A. Marshall Licht*)

saturated hydrocarbons or paraffins

Hydrocarbons are compounds that contain only carbon and hydrogen. Many hydrocarbons are known. These have been organized into several series or classes, according to similarities in their structures. One series of hydrocarbons is known as the alkanes. They also are called saturated hydrocarbons or paraffins. The term *saturation* means that each carbon atom is attached directly to the maximum number of other atoms. To visualize how this series can be established, consider first the smallest member of the series, *methane*. Methane is the principal component of the natural gas used for cooking and heating in many city homes. It consists of one carbon atom and four hydrogen atoms and has the structure shown.

It is possible to replace one of the hydrogen atoms in methane with a carbon atom. Since this new, attached carbon atom is capable of being bonded to four atoms, its three other positions will be bound to hydrogen.

$$
\begin{array}{ccc}
 & H & H \\
 & | & | \\
H— & C—C & —H \\
 & | & | \\
 & H & H
\end{array}
$$

Because the CH_3 group closely resembles a methane molecule, it is called a *methyl group*. The new compound contains two carbon atoms and six hydrogens and is called *ethane*.

The higher members of the alkane series of hydrocarbons are formed by repeating the substitution process. The next higher member after ethane is *propane*.

The next compound beyond propane is called *butane*.

Propane and butane are the principal components of the bottled gases that are used extensively in rural areas as fuel for cooking and heating.

The process of adding carbon atoms may be continued indefinitely, resulting in longer and longer hydrocarbon chains. Since each of the compounds formed in this manner will be linear, they often are called the **normal alkanes** or **straight-chain paraffins**.

With four or more carbon atoms, it is possible to have branched-chain paraffins in addition to the straight-chain paraffins. These compounds are isomers of the corresponding straight-chain paraffins.

There are two *butanes*.

$$
\begin{array}{c}
H \\
| \\
H—C—H \\
| \\
Methane
\end{array}
$$

$$
\begin{array}{c}
H \quad H \\
| \quad | \\
H—C—C—H \\
| \quad | \\
H \quad H \\
Ethane
\end{array}
$$

$$
\begin{array}{c}
H \quad H \quad H \\
| \quad | \quad | \\
H—C—C—C—H \\
| \quad | \quad | \\
H \quad H \quad H \\
Propane
\end{array}
$$

$$
\begin{array}{c}
H \quad H \quad H \quad H \\
| \quad | \quad | \quad | \\
H—C—C—C—C—H \\
| \quad | \quad | \quad | \\
H \quad H \quad H \quad H \\
Butane
\end{array}
$$

Normal butane
or *n*-butane

Isobutane

There are three *pentanes*.

```
          H
          |
      H—C—H
  H       |     H   H
  |       |     |   |
H—C———————C—————C———C—H
  |       |     |   |
  H       H     H   H
```

```
  H   H   H   H   H
  |   |   |   |   |
H—C———C———C———C———C—H
  |   |   |   |   |
  H   H   H   H   H
```

```
              H
              |
          H—C—H
  H           |       H
  |           |       |
H—C———————————C———————C—H
  |           |       |
  H           |       H
          H—C—H
              |
              H
```

There are five *hexanes*.

```
  H   H   H   H   H   H
  |   |   |   |   |   |
H—C———C———C———C———C———C—H
  |   |   |   |   |   |
  H   H   H   H   H   H
```

```
  H   H   H   H   H
  |   |   |   |   |
H—C———C———C———C———C—H
  |   |   |   |   |
  H   H   |   H   H
      H—C—H
          |
          H
```

```
              H
              |
          H—C—H
  H           |   H   H   H
  |           |   |   |   |
H—C———————————C———C———C———C—H
  |           |   |   |   |
  H           H   H   H   H
```

```
  H   H           H   H
  |   |           |   |
H—C———C———————————C———C—H
  |   |           |   |
  H   |           |   H
  H—C—H         H—C—H
      |             |
      H             H
```

```
                  H
                  |
              H—C—H
  H               |       H   H
  |               |       |   |
H—C———————————————C———————C———C—H
  |               |       |   |
  H               |       H   H
              H—C—H
                  |
                  H
```

Since the differences among the isomers are in their structures, they are known as **structural isomers.** Each structural isomer has different physical and chemical properties.

All hydrocarbons discussed to this point conform to the general formula C_nH_{2n+2}, where n is the number of carbon atoms in the molecule. It is possible to produce hydrocarbons that do not conform to this formula, however. For example, sometimes hydrocarbons form rings, as shown below.

Cyclopropane   Cyclobutane   Cyclopentane

Saturated hydrocarbons containing one ring have the general formula C_nH_{2n}.

As the structures become more complicated, it is increasingly more difficult to draw them in the manner previously described. Consequently, the symbols for carbon and hydrogen often are omitted. In these cases, it is assumed that there is a carbon atom at the junction of two or more bonds and that the appropriate number of hydrogen atoms is also present. Using this system, the *cyclopropane*, *cyclobutane*, and *cyclopentane* molecules just discussed may be represented as shown below.

Cyclopropane   Cyclobutane   Cyclopentane

physical properties of hydrocarbons

As the number of carbon atoms in a given series of compounds increases, there is a gradual transition in physical properties. The lower-molecular-weight hydrocarbons are gases with steadily increasing liquefaction points as carbon content increases. The intermediate-molecular-weight hydrocarbons are liquids with steadily increasing boiling points, and the higher-molecular-weight hydrocarbons are solids with steadily increasing melting points. Table 12-1 lists some physical constants for selected normal paraffins.

Table 12-1. Physical constants for selected normal paraffins

Name	Formula	Melting Point (°C)	Boiling Point (°C)	Uses
Methane	CH_4	−184	−161.5	Natural gas
Ethane	C_2H_6	−172	− 88.3	Petrochemical
Propane	C_3H_8	−189.9	− 42.2	Bottled gas
n-Butane	C_4H_{10}	−135	− 0.5	Bottled gas
n-Pentane	C_5H_{12}	−131.5	36.1	Constituents of
n-Hexane	C_6H_{14}	− 94.0	68.7	gasoline (among
n-Heptane	C_7H_{16}	− 90.5	98.4	others)
n-Octane	C_8H_{18}	− 56.5	125.8	Constituents of
n-Decane	$C_{10}H_{22}$	− 31	174	kerosene and diesel fuel
n-Pentadecane	$C_{15}H_{32}$	10	270.5	Constituents of
n-Eicosane	$C_{20}H_{42}$	38		lubricating oils
n-Pentacosane	$C_{25}H_{52}$	53.3		

The physical properties of the various hydrocarbons lead to different uses. The lower-molecular-weight hydrocarbons are gases and are used primarily as gaseous fuels. Methane, as mentioned earlier, is the natural gas used in city homes for cooking and heating. It is too expensive to run gas lines to homes that are widely scattered, as in rural areas. For these uses, it is desirable to transport and store gas as a liquid under pressure in tanks. Methane can be handled in this manner, but it requires a higher pressure for liquefaction than does propane or butane. Consequently, the latter gases, rather than methane, are used for household purposes in rural areas. Butane cannot be used during the winter in extremely cold climates because it liquifies at −0.5°C, or about the freezing point of water. At temperatures much below this level there is not sufficient gas pressure to operate stoves and heaters.

The longer-chain hydrocarbons are used for gasoline, kerosene, diesel fuel, and heating oil. Hydrocarbons with even longer chains find use in lubricating oils and waxes.

unsaturated hydrocarbons

In the discussion of inorganic complex ions and molecules (Chapter 8), it was noted that in some of these species double and even triple covalent bonds occur. This situation also holds for hydrocarbons.

The smallest hydrocarbon that can contain a double bond between two carbon atoms is *ethylene*.

$$\begin{array}{ccc} H & & H \\ \diagdown & & \diagup \\ & C{=}C & \\ \diagup & & \diagdown \\ H & & H \end{array}$$

Ethylene

Single bonds allow free rotation of groups around the bonds. This is not true for double bonds. Atoms cannot rotate around double bonds. This leads to another type of isomerism known as **geometric isomerism**. Consider, for example, the unsaturated hydrocarbon called 2-*butene* (the 2 indicates that the double bond is on the second carbon from the end). There are two kinds of 2-butene, resulting from geometric isomerism.

$$\begin{array}{ccc} H_3C & & CH_3 \\ \diagdown & & \diagup \\ & C{=}C & \\ \diagup & & \diagdown \\ H & & H \end{array} \qquad\qquad \begin{array}{ccc} H_3C & & H \\ \diagdown & & \diagup \\ & C{=}C & \\ \diagup & & \diagdown \\ H & & CH_3 \end{array}$$

cis-2-butene   *trans*-2-butene

When both CH_3 groups are on the same side of the double bond the molecule is said to have a *cis* configuration. When the CH_3 groups are on opposite sides of the double bond, the molecule is said to have a *trans* configuration. This difference in geometry leads to different physical and chemical properties for the two types of molecules. Natural rubber, for example, is a large molecule with only *cis* double bonds. One of the original synthetic rubbers consisted of the same molecule with a random distribution of *cis* and *trans* double bonds. The natural form of rubber generally is considered superior to this form of synthetic rubber in wear and stretch properties.

Hydrocarbons containing double bonds are known as **alkenes** or **olefins**. A series of olefins containing one double bond may be derived in a manner similar to that of the paraffins' derivation. The position of the double bond may be varied, as well as the number of carbons in the chain. Olefins containing one double bond correspond to the general formula C_nH_{2n}, where n is an integer. Figure 12-2 shows the structures of a few olefins.

Hydrocarbons containing triple bonds are known as *alkynes*. The smallest alkyne is *acetylene*, which is used extensively in industry as a chemical raw material and as a welding fuel.

$$H{-}C{\equiv}C{-}H$$

Acetylene

$CH_3 - CH = CH_2$ propylene	H_3C $\quad\quad C = CH_2$ H_3C isobutylene	cyclohexene
$CH_3 - CH = CH - CH_2 - CH_3$ 2-pentene		

Figure 12-2   Some examples of olefins

aromatic hydrocarbons

The aromatic hydrocarbons are so called because many of them, as well as their chemical derivatives, have rather pleasant odors. Their unique structural and chemical properties put them in a special class.

The smallest and most common member of the series is *benzene* (C_6H_6). Benzene consists of a six-membered ring containing three alternating double bonds. The simple picture of three alternating single and double bonds does not explain adequately benzene's chemical properties, however. It is impossible to distinguish between the single and double bonds of benzene. In fact, benzene appears to be a hybrid, called a *resonance hybrid*, between the two possible structures.

These two structures are known as the *Kekulé* structures of benzene, after Friedrich Kekulé who first proposed them in 1865. The double-headed arrow normally designates resonance hybrids. To better represent the structural and chemical characteristics of aromatic compounds, the benzene ring is now often represented as the following.

A large number of compounds contain the basic ring structure of benzene. Some of the more common are shown in Figure 12-3. Some compounds contain two or more benzene rings fused together,

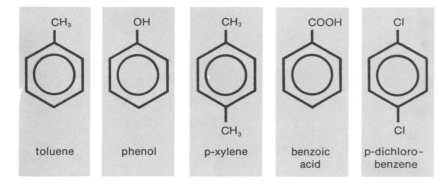

Figure 12-3. Some common compounds containing the benzene ring

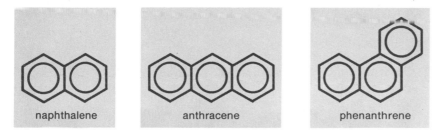

Figure 12-4. Some fused benzene compounds

as illustrated in Figure 12-4. Of the compounds shown, *naphthalene* is probably best known because of its use in mothballs. The compound *p-dichlorobenzene* (Figure 12-3) also is used in mothballs.

petroleum

The principal source of hydrocarbons for industrial use is **petroleum**. Petroleum is a foul-smelling, dark, oily substance that is found in underground deposits in various parts of the world. It is primarily a mixture of hydrocarbons, but its exact composition varies from one location to another.

In its natural or crude state, petroleum is largely useless. It may be refined or converted to useful products relatively easily, however. The most important step in the refining process is **fractional distillation**. In the fractional distillation of petroleum, the crude substance is heated until the volatile components are vaporized. The vapors then pass through a fractionating tower, which contains

A petroleum refinery and petrochemical complex's tall cracking and fractionating towers, storage tanks, and transportation facilities. (*Courtesy Peter Stackpole, Exxon Company, USA*)

Figure 12-5. A simplified petroleum fractionating tower and some of the purified products

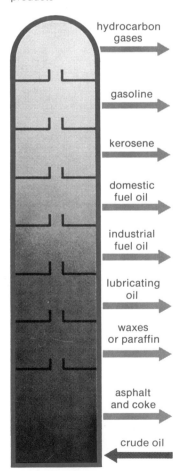

hydrocarbon gases

gasoline

kerosene

domestic fuel oil

industrial fuel oil

lubricating oil

waxes or paraffin

asphalt and coke

crude oil

several compartments at successively lower temperatures. Thus the vapors cool as they pass through the tower. As the vapors cool, they condense (liquefy). The higher-boiling vapors condense in the lower and warmer portions of the fractionating tower, and the lower-boiling vapors condense in the upper and cooler portions. The various fractions are withdrawn from the tower for further purification. Figure 12-5 is a drawing of a fractionating tower.

Gasoline, which constitutes only a relatively small portion of the normal distillates, is the major commercial product from petroleum. Consequently, it is necessary to convert some of the other hydrocarbon fractions into gasoline. The main process for converting higher-boiling hydrocarbons into gasoline is known as catalytic cracking. In the process, the hydrocarbon to be cracked is vaporized, and the vapors are passed over a hot bed of catalyst. (A catalyst is a substance that affects the rate of a chemical reaction, but is not permanently changed by the reaction.) On the catalyst, the hydrocarbon molecules are cleaved into smaller molecules. By the proper choice of catalyst and operating conditions, it is possible to optimize the amount of gasoline that can be obtained from the catalytic cracker.

A related process, *catalytic reforming*, is used to upgrade the quality of gasoline by catalytically converting linear hydrocarbons to aromatic ones.

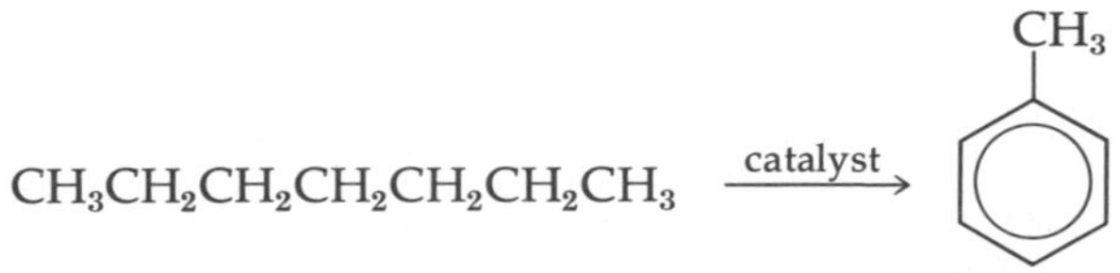

$$CH_3CH_2CH_2CH_2CH_2CH_2CH_3 \xrightarrow{\text{catalyst}}$$

n-heptane

A low-quality gasoline
component

toluene

A high-quality gasoline
component

Aromatic hydrocarbons have better combustion characteristics in the automobile engine than linear hydrocarbons, as discussed in Chapter 17.

In addition to its use as a fuel, petroleum is the principal source of hydrocarbon raw materials for the chemical industry. Plastics, textile fibers, detergents, and paint thinners are just a few of the common materials derived from petroleum.

functional groups

Carbon can form covalent bonds with a number of other elements besides hydrogen. Of these, the halogens, oxygen, nitrogen, and sulfur, are most important. Oxygen in particular makes several kinds of compounds with carbon. The various groups that carbon forms with these elements are called functional groups. Different types of organic compounds result when the functional groups are attached to a hydrocarbon skeleton. Functional groups are important because they are frequently sites of chemical activity. Table 12-2 lists some common functional groups and some important compounds of each type.

alcohols and ethers

Structurally, *alcohols* and *ethers* may be considered organic derivatives of water. They are obtained by replacing the hydrogen atom in water with hydrocarbon groups.

$$H—O—H \qquad CH_3CH_2—O—H \qquad CH_3CH_2—O—CH_2CH_3$$

Water

Ethanol
(Ethyl alcohol)

Ethyl ether

Table 12-2. Typical functional groups and related compounds

Functional Group		Typical Compound		
Structure	Group Name	Structure	Name	Uses or Common Names
$-OH$	Alcohol	CH_3OH	Methyl alcohol	Wood alcohol (poisonous)
$-O-$	Ether	$CH_3CH_2OCH_2CH_3$	Diethyl ether	Anesthetic and industrial solvent
$\overset{O}{\overset{\|}{-C-H}}$	Aldehyde	$H-\overset{O}{\overset{\|}{C}}-H$	Formaldehyde	Formalin (40% HCHO in water)
		$CH_3\overset{O}{\overset{\|}{C}}-H$	Acetaldehyde	Odor of apples
$\overset{O}{\overset{\|}{-C-}}$	Ketone	$CH_3\overset{O}{\overset{\|}{C}}-CH_3$	Acetone	Industrial solvent
$\overset{O}{\overset{\|}{-C-OH}}$	Carboxylic acid	$CH_3\overset{O}{\overset{\|}{C}}-OH$	Acetic acid	Active component of vinegar
		benzoic acid structure $-\overset{O}{\overset{\|}{C}}-OH$	Benzoic acid	Sodium benzoate is a common food preservative.
$\overset{O}{\overset{\|}{-C-O-}}$	Esters	$CH_3\overset{O}{\overset{\|}{C}}-OCH_2CH_3$	Ethyl acetate	Lacquer thinner
$\overset{O}{\overset{\|}{-C-N-}}$	Amide	caprolactam ring $N-H$	Caprolactam	Starting material for Nylon 6
$-NH_2$	Primary amine	$CH_3CH_2NH_2$	Ethyl amine	Amines are generally responsible for fish odors
$-NH$	Secondary amine	$(CH_3CH_2)_2NH$	Diethyl amine	
$-N-$	Tertiary amine	$(CH_3CH_2)_3N$	Triethyl amine	
$-N^{\pm}-$ X^-	Quaternary ammonium salt	$(CH_3CH_2)_4N^+Cl^-$	Tetraethyl ammonium chloride	Some quaternary ammonium salts are used as antiseptics
$-Cl$ $(-Br)$ $(-F)$ $(-I)$	Halide	$CHCl_3$	Chloroform	Industrial solvent and formerly an anesthetic
		$Cl_2C=CCl_2$	Tetrachloroethylene	Dry cleaning solvent

The alcohol with the simplest structure is *methyl alcohol,* or *methanol.* It is sometimes called *wood alcohol* because it was obtained originally from wood. It is poisonous, causing blindness or death if taken internally. The industrial route to methanol involves the reaction of carbon monoxide and hydrogen.

$$CO + 2H_2 \rightarrow CH_3OH$$
Methanol

It is used in large quantities as an industrial solvent and reagent.

Ethyl alcohol, or *ethanol,* is probably the best known alcohol because it is the one contained in alcoholic beverages. Owing to federal law, all alcohol used in alcoholic beverages, foods, and drugs is made by fermenting grains with yeast. It is more economical, however, to make ethanol by adding water to *ethylene,* using sulfuric acid as a catalyst.

$$CH_2\!\!=\!\!CH_2 + H_2O \xrightarrow{\ H_2SO_4\ } CH_3CH_2OH$$
Ethylene   Ethanol

For industrial use, ethanol is usually made this way. Much industrial ethanol is denatured. *Denatured alcohol* is ethanol that has been treated with methanol or some other poison, making the ethanol unfit to drink.

Another common alcohol is *isopropyl alcohol,* or *isopropanol.* A mixture of 70% isopropanol and water is sold in drug stores as "rubbing alcohol." It has an antiseptic quality and leaves a pleasant, cooling sensation as it evaporates from the skin. Isopropanol is poisonous, however, and should not be taken internally. It is prepared by adding water to *propylene*

$$CH_3CH\!\!=\!\!CH_2 + H_2O \rightarrow CH_3\underset{\underset{\textstyle OH}{|}}{C}HCH_3$$

Propylene   Isopropanol

or by adding hydrogen to *acetone.*

$$\underset{\textstyle CH_3CCH_3}{\overset{\textstyle \overset{O}{\|}}{}} + H_2 \rightarrow \underset{\textstyle CH_3CHCH_3}{\overset{\textstyle \overset{OH}{|}}{}}$$

Acetone

Glycerine, or *glycerol,* is an alcohol that contains three hydroxyl (OH) groups.

$$\underset{\textstyle OH\ \ \ OH\ \ \ OH}{\overset{\textstyle CH_2\!-\!CH\!-\!CH_2}{|\ \ \ \ \ |\ \ \ \ \ |}}$$
Glycerol

It is an important constituent of fats and is obtained as a by-product in soap manufacture.

$$\text{fat} + \text{lye} \rightarrow \text{soap} + \text{glycerol}$$

The process of making soap from fat is discussed in more detail in Chapter 15.

The only well-known ether is *ethyl ether*, which is used as an anesthetic and as an industrial and laboratory solvent. It is made along with ethanol from ethylene.

$$CH_2{=}CH_2 + H_2O \xrightarrow{\ H_2SO_4\ } CH_3CH_2OCH_2CH_3$$

Ethylene   Ethyl ether

$$+ CH_3CH_2OH$$

Ethanol

The product ratio is varied to accommodate the commercial demand by adjusting the temperature and sulfuric acid concentration.

aldehydes and ketones

Aldehydes and *ketones* are structurally similar types of compounds. The only difference is the position of the C=O group on the molecule. The C=O group is called a carbonyl group. When the carbonyl group is on the end of a carbon chain, the molecule is a ketone.

$$CH_3CH_2{-}C{\Large\langle}^{O}_{H} \qquad\qquad CH_3{-}\overset{\overset{\textstyle O}{\|}}{C}{-}CH_3$$

An aldehyde   A ketone

Formaldehyde

Compounds containing aldehyde and ketone groups are rather common. *Formaldehyde* is probably the best-known aldehyde. It is used in biology and medicine as a preservative.

Acetone is probably the most common ketone. It is used extensively as an industrial solvent for plastics. It may be made by oxidation of isopropanol.

$$\overset{\overset{\textstyle OH}{|}}{CH_3CHCH_3} \xrightarrow{\ [O]\ } \overset{\overset{\textstyle O}{\|}}{CH_3CCH_3} + H_2O$$

Isopropanol   Acetone

The symbol [O] over the arrow in the equation indicates that this is an oxidation reaction, but does not specify what oxidizing agent is used. Many oxidizing agents may be used in this reaction.

carboxylic acids, esters, and amides

Like aldehydes and ketones, *carboxylic acids*, *esters*, and *amides* are structurally similar compounds. The difference is in the group attached to the carbonyl group.

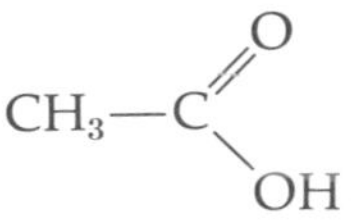

A carboxylic acid 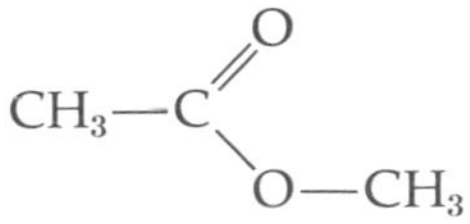   An ester 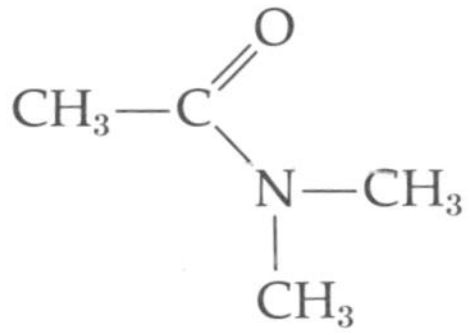   An amide

Acetic acid is probably the most common organic acid. Vinegar is a solution of about 5% acetic acid in water. It is obtained by the excessive fermentation of apple cider. The conversion of ethanol into acetic acid in this process is accomplished by bacteria.

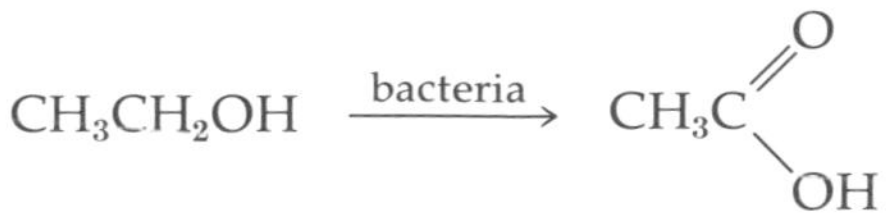

$$CH_3CH_2OH \xrightarrow{\text{bacteria}} CH_3C\overset{O}{\underset{OH}{<}}$$

Ethanol   Acetic acid

There are several industrial routes to acetic acid. One of the more recently developed processes involves the condensation of carbon monoxide with methanol.

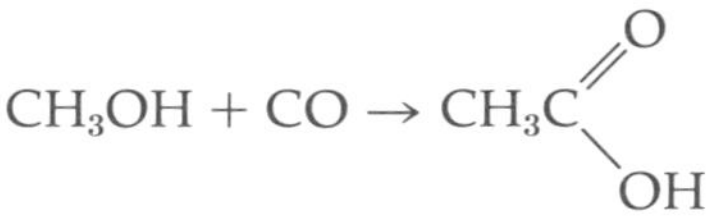

$$CH_3OH + CO \rightarrow CH_3C\overset{O}{\underset{OH}{<}}$$

Methanol   Acetic acid

Although most people do not know it by name, *ethyl acetate* is probably one of the most common industrial esters. It is used extensively as a lacquer thinner. Like most esters, it is made by treating a carboxylic acid (acetic acid) with an alcohol (ethanol) in the presence of a mineral acid (usually sulfuric acid or phosphoric acid). A by-product of the reaction is water.

$$CH_3\overset{O}{\overset{\|}{C}}OH + CH_3CH_2OH \xrightarrow{H_2SO_4} CH_3\overset{O}{\overset{\|}{C}}{-}OCH_2CH_3 + H_2O$$

Acetic acid   Ethanol   Ethyl acetate

Other esters are prepared by using different carboxylic acids or alcohols.

Esters are usually pleasant-smelling compounds with fruitlike odors. In fact, specific esters are responsible for the odor of such plant materials as wintergreen and bananas. Table 12-3 lists the esters associated with some common odors and flavors.

Table 12-3. Ester components of some common odors and flavors

Isoamyl acetate	bananas
Isobutyl propionate	rum
n-Butyl butyrate	pineapples
n-Amyl butyrate	apricots
Isoamyl valerate	apples
Octyl acetate	oranges
Benzyl acetate	jasmine
Methyl salicylate	wintergreen

Amides may be made by treating carboxylic acids with ammonia or amines and heating the resulting salt.

$$CH_3\overset{O}{\overset{\diagup\!\!\diagdown}{C}}{\diagdown}_{OH} + NH_3 \rightarrow CH_3\overset{O}{\overset{\diagup\!\!\diagdown}{C}}{\diagdown}_{O^{-1}} NH_4^{+1} \rightarrow CH_3\overset{O}{\overset{\diagup\!\!\diagdown}{C}}{\diagdown}_{NH_2} + H_2O$$

Acetic acid   Ammonium   Acetamide
                    acetate

Proteins and nylon are common substances that contain amide groups. Proteins are discussed in Chapter 13 and nylon in Chapter 16.

amines

Amines may be thought of as organic derivations of ammonia. They are obtained by replacing the hydrogens in ammonia with hydrocarbon groups.

$$\begin{array}{c} H \\ | \\ H-N-H \\ \text{Ammonia} \end{array} \qquad \begin{array}{c} H \\ | \\ CH_3-N-H \\ \text{Primary} \\ \text{amine} \end{array} \qquad \begin{array}{c} CH_3 \\ | \\ CH_3-N-H \\ \text{Secondary} \\ \text{amine} \end{array} \qquad \begin{array}{c} CH_3 \\ | \\ CH_3-N-CH_3 \\ \text{Tertiary} \\ \text{amine} \end{array}$$

Amines are responsible for the odor normally associated with fish.

Like ammonia, which can react with acids to form ammonium salts, amines can react with acids to form amine salts. For example, ammonia will react with hydrochloric acid to form ammonium chloride.

$$NH_3 + HCl \rightarrow NH_4{}^{+1}Cl^{-1}$$

Similarly, trimethylamine will react with hydrochloric acid to form trimethylammonium chloride.

$$(CH_3)_3N + HCl \rightarrow (CH_3)_3NH^{+1}Cl^{-1}$$

Trimethylamine   Trimethylammonium
                        chloride

A related class of compounds is the quaternary amine salts. These compounds have four hydrocarbon residues attached to the nitrogen.

$$\begin{array}{c} CH_3 \\ |+1 \\ CH_3-N-CH_3 \qquad I^{-1} \\ | \\ CH_3 \end{array}$$

Quaternary salts have antiseptic properties and are used sometimes in place of swimming pool chlorine.

halides

When one or more of the hydrogen atoms in an organic molecule is replaced with a halogen (fluorine, chlorine, bromine, or iodine), the resulting compound is called an *organic halide*. The most common compounds are the organic chlorides. Compounds like *tetrachloroethylene* ($Cl_2C{=}CCl_2$) are used extensively as solvents in dry-cleaning clothes. *Carbon tetrachloride* (CCl_4) was once sold as a spot remover for home use, but has since been taken off the market because it is believed to cause liver damage.

Selected organic fluorine compounds are used as refrigerants in home refrigerators under the trade name of Freon. *Dichlorodifluoromethane* (CCl_2F_2), commonly called Freon-12, is used most extensively in home refrigerators. Freon also is used as the propellant in aerosol cans.

organic reactions

It is impossible to cover all organic reactions or organic compounds in only one chapter. Some of the more important and characteristic reactions of the major types of organic compounds are given in Table 12-4. The R attached to the functional groups commonly represents any hydrocarbon group. When a second, different R group is attached to the molecule, the symbol R' is used.

Organic compounds have important industrial functions as solvents and as reactants for the manufacture of such items as synthetic fibers, plastics, medicinal compounds, and petroleum prod-

Table 12-4. Some common organic reactions

$$RCH_2OH \xrightarrow{[O]} RCHO \xrightarrow{[O]} RCOOH$$

Alcohol   Aldehyde   Acid

$$R_2CHOH \xrightarrow{[O]} R_2C{=}O$$

Alcohol   Ketone

$$ROH \xrightarrow{HBr} RBr$$

Alcohol   Alkyl bromide

$$RC{\overset{O}{\underset{OH}{}}} + R'OH \rightarrow RC{\overset{O}{\underset{OR'}{}}} + H_2O$$

Acid   Alcohol   Ester

$$RCH{=}CH_2 \xrightarrow{HOH} RCH{-}CH_3$$
$$\underset{OH}{|}$$

Olefin   Alcohol

$$RC{\overset{O}{\underset{OH}{}}} + R'NH_2 \xrightarrow{\Delta} RC{\overset{O}{\underset{NHR'}{}}} + H_2O$$

Acid   Amine   Amide

ucts. Much material found in living matter also is composed of organic compounds, and the life processes that occur in living cells are primarily organic reactions. In the next few chapters, some specific applications of organic chemistry and organic reactions in the areas of biochemistry, medicinal chemistry, and industry will be discussed.

study questions

1 Compare the original and modern definitions of organic chemistry.

2 Why is carbon an unusually important element?

3 Draw the structure of an optically active compound.

4 By what experimental technique are two optical isomers (mirror images) distinguished from one another?

5 Name the four major classes of hydrocarbons.

6 What difficulty is encountered in using bottled butane as a fuel in cold climates?

7 Distinguish between structural and geometrical isomerism.

8 Give an example of an alkane, an alkene, and an alkyne, each containing four carbon atoms.

9 How is gasoline manufactured?

10 How can the yield of gasoline from petroleum be increased?

11 What materials other than gasoline are obtained from petroleum?

12 A shortage of petroleum creates problems not only in gasoline supplies but also in the manufacture of a large number of commercial products. Why?

13 Identify the functional class (alcohol, acid, and so on) to which each of the following compounds belongs.

a. $CH_3-CH_2-\overset{\overset{\displaystyle O}{\|}}{C}-H$

b. $CH_3-CH_2-\underset{\underset{\displaystyle OH}{|}}{CH}-CH_3$

c. benzene ring $-CH_2-O-CH_3$

d. $CH_3-\underset{\underset{\displaystyle CH_3}{|}}{CH}-\overset{\overset{\displaystyle O}{\|}}{C}-OH$

e. $CH_3-CH_2-NH-CH_3$

f. $CH_3-\overset{\overset{\displaystyle O}{\|}}{C}-O-$ benzene ring

g. $CH_3-CH_2-\overset{\overset{\displaystyle O}{\|}}{C}-CH_3$

special problems

1 Draw all the isomers of C_5H_{12}.

2 Write a balanced equation for the combustion of butane in air to carbon dioxide and water.

3 By referring to Table 12-3, show how the following conversions might be accomplished.

a. Ethyl alcohol to acetic acid

b. CH_3—CH=CH—CH_3 to CH_3—CH_2—$\underset{\underset{Br}{|}}{CH}$—$CH_3$

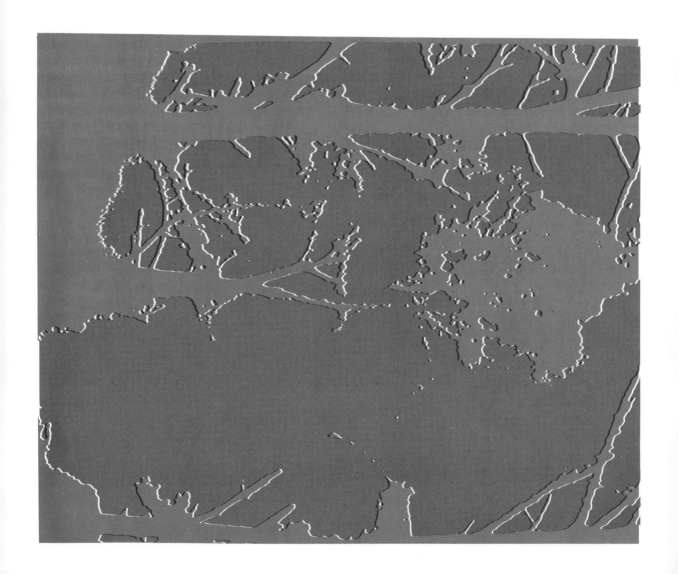

biochemistry

Biochemistry is the study of the chemistry of living systems. In earlier years this study consisted mainly of isolating and identifying the many complex molecules found in living cells. In recent years, however, significant strides have been made toward understanding the functions and interactions of these molecules during the normal operation of the cell.

This chapter provides a brief introduction to biochemistry. The principal biomolecules—carbohydrates, lipids, proteins, and nucleic acids—are described first, followed by a discussion of protein synthesis and cell replication.

carbohydrates

Carbohydrates are compounds that usually contain the elements carbon, hydrogen, and oxygen in the ratio 1 to 2 to 1. They conform to the general formula $(CH_2O)_n$, making them appear to be hydrates of carbon. Actually they are much more complex than their name would imply. Carbohydrates vary in size from small, simple molecules like glyceraldehyde (monosaccharides) to large, complex polymers like starch and cellulose (polysaccharides).

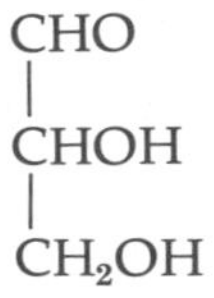

carbohydrate
enzyme
photosynthesis
glycolysis
tricarboxylic acid cycle
fatty acid
lipid
β oxidation
protein
amino acid
peptide linkage
nucleic acid
deoxyribonucleic acid (DNA)
ribonucleic acid (RNA)
double helix
ribosome
codon
genetic code

CHO
|
CHOH
|
CH$_2$OH

Glyceraldehyde

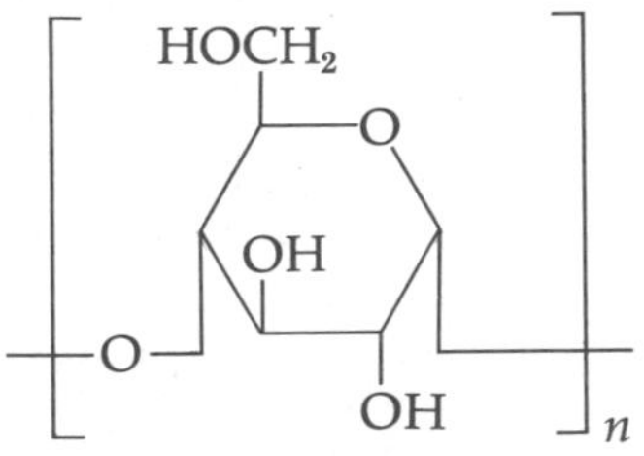

Starch

hexoses

Sugars, which are the relatively simple carbohydrates, are polyhydroxy aldehydes or ketones. Probably the most common sugars are the hexoses, which have the general formula $C_6H_{12}O_6$. Of these, *glucose* and *fructose* are best known.

Glucose is the most abundant sugar found in nature. Not only is it the principal fuel for most organisms, but also it is the building block for the most important and naturally occurring polysaccharides. Glucose belongs to a group of hexose sugars known as the aldoses because it is a polyhydroxy aldehyde.

Fructose is a ketose because it is a polyhydroxy ketone. In both glucose and fructose, each of the remaining carbon atoms has a hydroxyl group attached to it.

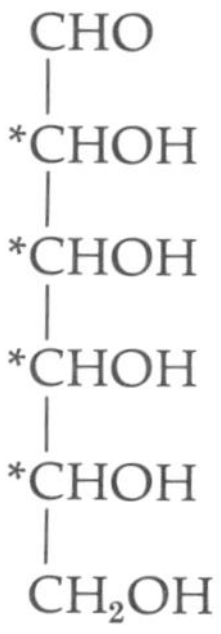

CHO   CH₂OH

| |
*CHOH   C=O
| |
*CHOH   *CHOH
| |
*CHOH   *CHOH
| |
*CHOH   *CHOH
| |
CH₂OH   CH₂OH

Glucose 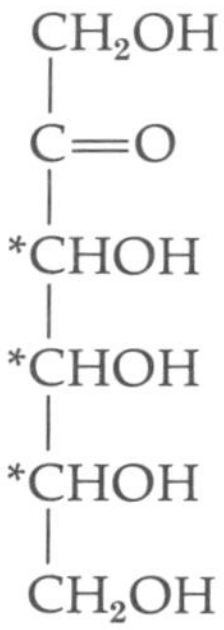   Fructose

(The asterisked carbons are asymmetric and are sites of optical activity in these molecules.)

Most hexoses do not exist in the simple open-chain structures shown. Instead, the aldehyde or ketone group tends to react with one of the hydroxyl groups on the chain to form a cyclic compound. Glucose forms a six-membered ring containing one oxygen, and fructose forms a five-membered ring containing one oxygen.

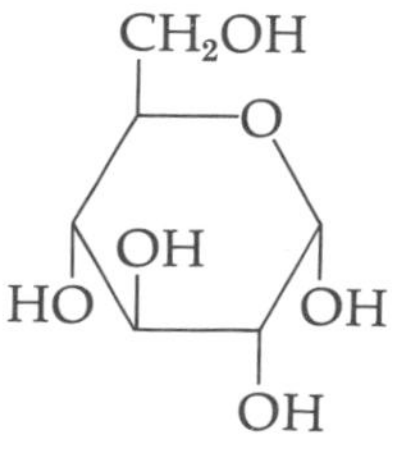

Glucose 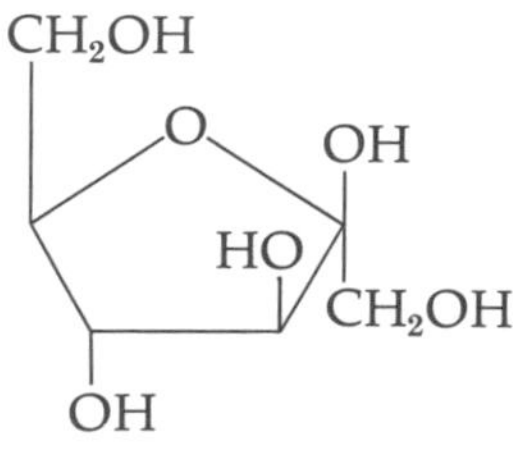   Fructose

Dextrose, the common naturally occurring form of glucose, rotates plane-polarized light to the right (dextrorotatory), and levulose, the common naturally occurring form of fructose, rotates plane-polarized light to the left (levorotatory). The rotation of plane-polarized light is due to the presence of optically active centers in these molecules.

disaccharides

The simple sugars, called *monosaccharides*, can link together to form more complex members of the carbohydrate family. Disaccharides, for example, are formed from two simple sugar units, trisaccharides from three units, and polysaccharides from many units. The bonding of monosaccharide units results from the chemical interaction of a hydroxyl group on one sugar molecule with a hydroxyl group on another sugar molecule, to form an etherlike linkage.

Sucrose, the common table sugar found in grocery stores and used in cooking, is a disaccharide of glucose and fructose.

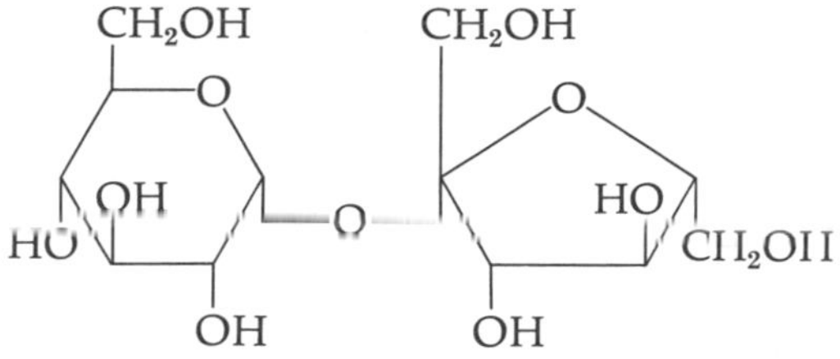

Sucrose

This compound is obtained principally from sugar cane or sugar beets and is responsible for a major industry.

Two other common disaccharides are *lactose* and *maltose*.

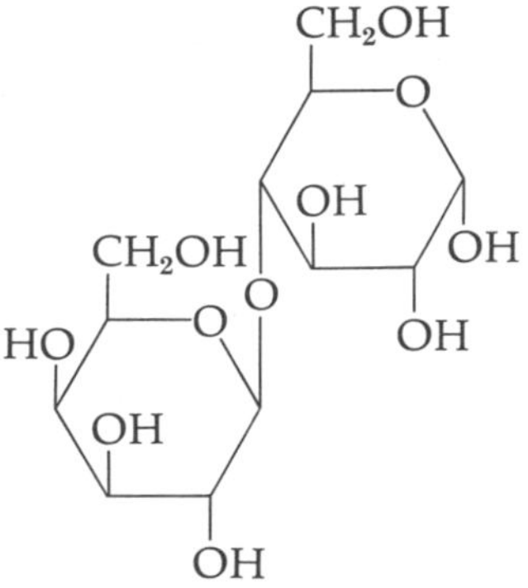

Lactose

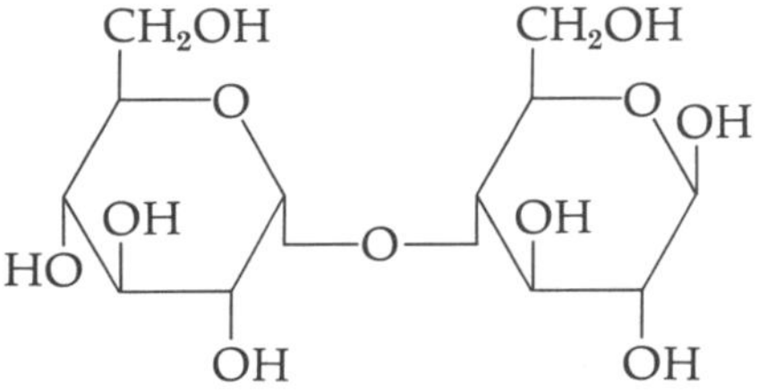

Maltose

Lactose is a disaccharide made by combining glucose and *galactose*; it is most common in milk. Maltose consists of two glucose molecules connected together. It usually is obtained by the partial hydrolysis of starch.

polysaccharides

Starch and *glycogen*, which are polysaccharides, serve as reservoirs of sugar in plants and animals respectively. Starch and glycogen are large polymers, composed solely of glucose units. Starch contains two components, *amylose* and *amylopectin*. Amylose is a linear unbranched compound, but amylopectin is highly branched. Glycogen, the storage polysaccharide found in animals, also is branched, but more highly so than amylopectin.

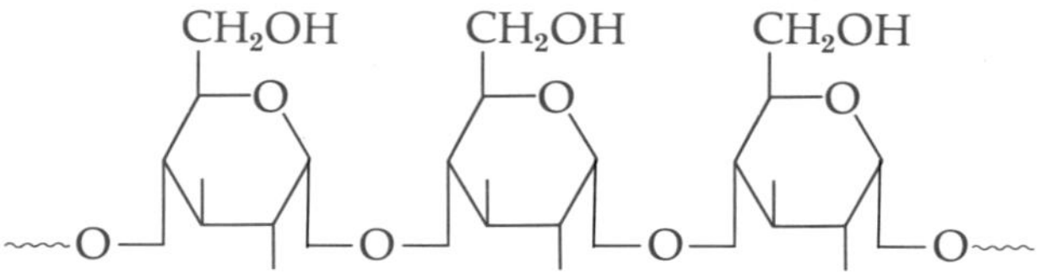

Amylose

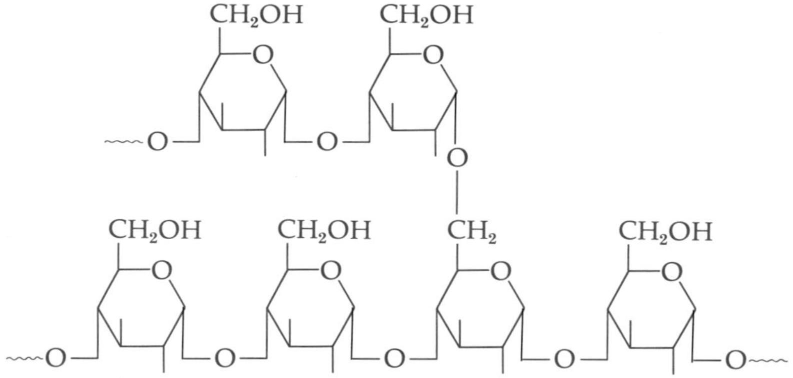

Glycogen

When glucose is present in excess, as frequently occurs in animals shortly after a large meal, the glucose is incorporated into starch and glycogen, usually by extending existing molecules of these polysaccharides.

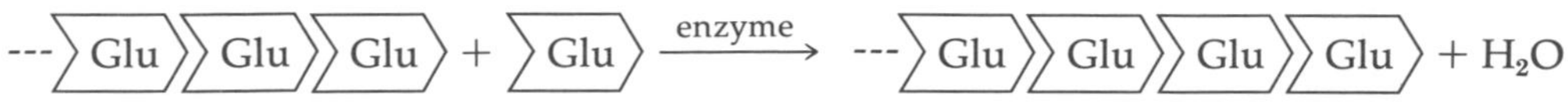

$$\text{---} \rangle \text{Glu} \rangle\rangle \text{Glu} \rangle\rangle \text{Glu} \rangle + \rangle \text{Glu} \rangle \xrightarrow{\text{enzyme}} \text{---} \rangle \text{Glu} \rangle\rangle \text{Glu} \rangle\rangle \text{Glu} \rangle\rangle \text{Glu} \rangle + H_2O$$

Partial structure of glycogen or starch molecule   Extended glycogen or starch molecule

When glucose is in short supply in the body, such as after strenuous exercise, glucose is released by partial degradation of glycogen.

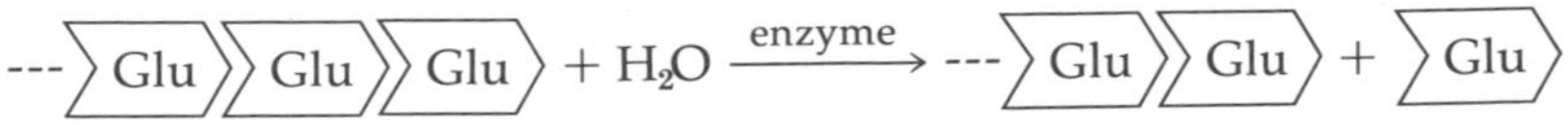

Glycogen

The relative rates of these two reactions are controlled by complex protein molecules known as enzymes. Enzymes control the rate and three-dimensional characteristics of biological reactions. Presently it is believed that there is a unique enzyme for almost every biochemical reaction.

Cellulose, another important polysaccharide, is also a polymer of glucose. In this case, however, the spatial orientation of the bonds that link the carbon atoms to the next sugar units (shown in color in the structure below) is different from the orientation of similar bonds found in starch and glycogen (shown in color in the earlier structures of amylose and glycogen). In cellulose, these bonds are oriented up, but in starch and glycogen, they are oriented down. This small, but significant, change in structure makes cellulose un-

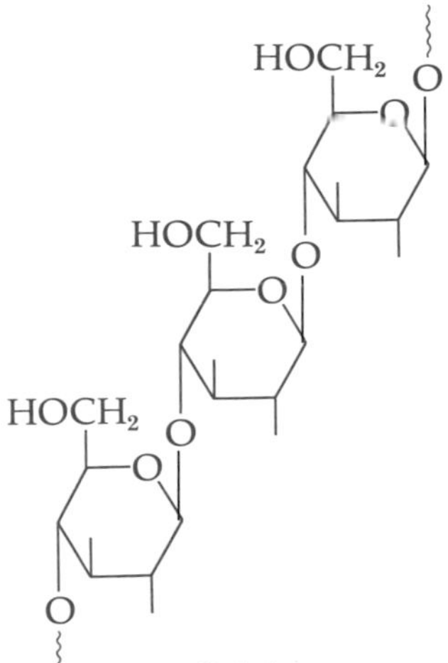

Cellulose

reactive in most animal and plant systems. For this reason, cellulose is an excellent structural component of plant cells and is principally responsible for the strength of wood.

The conversion of cellulose to glucose is theoretically possible, and some microorganisms actually have this capability.

$$(\text{glucose})_n + n\ H_2O \rightarrow n\ \text{glucose}$$

Cellulose

Some animals, such as cattle, can feed on cellulose because micro-organisms in their digestive systems can convert cellulose to glucose enzymatically. Most animals, however, are unable to break down cellulose and therefore cannot use this polysaccharide as food. If a commercial process for the conversion of cellulose to glucose could be developed, sawdust and other scrap lumber or paper products could be used as a source of food. Furthermore, an industrial waste material could be converted into a valuable commercial product. Laboratory-scale hydrolysis of cellulose has been achieved, but an economic large-scale process has yet to be developed.

Glucose is produced in green plants from water and atmospheric carbon dioxide. The process, called photosynthesis, requires light and involves a complex series of reactions.

$$6\ CO_2 + 6\ H_2O \xrightarrow{\text{sunlight}} C_6H_{12}O_6 + 6\ O_2$$
$$\text{Glucose}$$

This process is not spontaneous, because it requires energy to proceed. Light from the sun serves as the source of this energy. In photosynthesis, energy from the sun is utilized by the green leaves of plants to form carbohydrates and oxygen from carbon dioxide and water. The carbohydrates and oxygen are then consumed by animals who convert them back to carbon dioxide and water. The energy that is released in the process is used for such activities as walking and running. Figure 13-1 illustrates the cycle of photosynthesis.

Figure 13-1. The photosynthetic cycle in nature

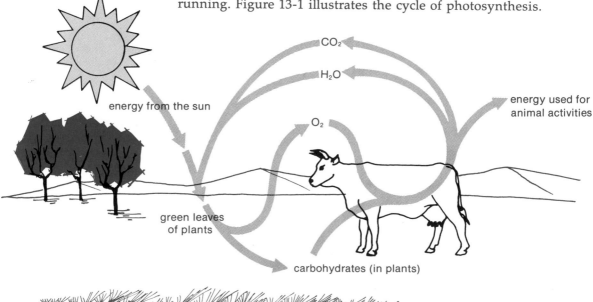

energy from the sun

CO_2

H_2O

O_2

energy used for animal activities

green leaves of plants

carbohydrates (in plants)

cellular energy generation

Energy is required for muscular and other body activities. It is believed currently that the immediate source of energy in living cells is the compound adenosine triphosphate, which commonly is abbreviated ATP.

Adenosine triphosphate (ATP)

Adenosine triphosphate is an energy storage compound formed from adenosine diphosphate (ADP) in certain metabolic processes in cells, such as the metabolic degradation of sugars. When energy is needed, as in the contraction of a muscle, ATP is reconverted to ADP, thereby releasing energy for use by the cells.

$$\text{ATP} + \text{H}_2\text{O} \xrightarrow{\text{enzyme}} \text{ADP} + \text{H}_3\text{PO}_4 + \text{energy}$$

The metabolic breakdown of sugars in living systems takes place by a series of chemical reactions. The most important of these reactions can be subdivided into two successive sets of processes known as **glycolysis** and the **tricarboxylic acid cycle**.

The series of reactions in glycolysis results in the breakdown of glucose to pyruvic acid. Each glucose molecule produces two molecules of pyruvic acid. All other hexose sugars in the living cell are converted to glucose by an associated sequence of reactions. An abbreviated scheme showing the fate of the carbon skeleton of glucose in glycolysis is given in Figure 13-2.

The sequence of reactions involves numerous enzymes and co-reactants. As a consequence of the degradation of glucose to pyruvic acid, there is a net conversion of 12 molecules of ADP to ATP for each molecule of sugar that is converted to pyruvic acid.

The tricarboxylic acid cycle is summarized in Figure 13-3. In this cycle, pyruvic acid is converted to carbon dioxide and water. The

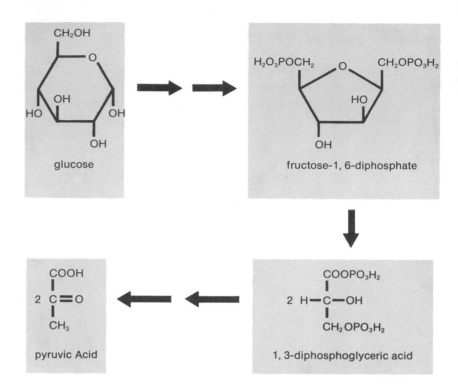

Figure 13-2 An abbreviated form of the glycolysis scheme

reaction course is called a *cycle* because the starting compound, oxaloacetic acid, is regenerated at the end of the cycle. As a consequence of the tricarboxylic acid cycle, 12 molecules of ATP are generated per molecule of pyruvic acid consumed. Therefore, as the result of both glycolysis and the tricarboxylic acid cycle, each molecule of glucose leads to the formation of 36 molecules of ATP, as glucose is converted to carbon dioxide and water.

$$\text{glucose} \rightarrow 2(\text{pyruvic acid}) \quad 12 \text{ ATP produced}$$

$$2(\text{pyruvic acid}) \rightarrow \begin{array}{c}\text{carbon dioxide}\\ \text{and water}\end{array} \quad 2(12 \text{ ATP}) \text{ produced}$$

Overall reaction: $\quad \text{glucose} \rightarrow \begin{array}{c}\text{carbon dioxide}\\ \text{and water}\end{array} \quad 36 \text{ ATP produced}$

The ATP molecules formed in these processes are used by the cell as energy sources for other metabolic functions.

The tricarboxylic acid cycle is a particularly interesting series of reactions, because it demonstrates some of the intricate interrelationships of chemical processes in the living system. For example, before

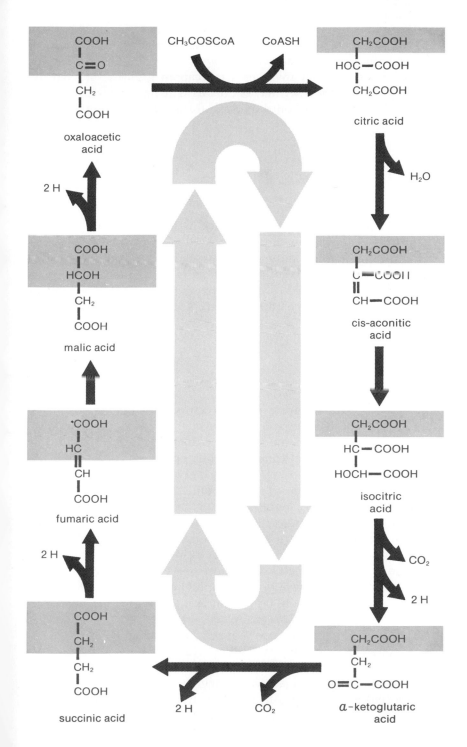

Figure 13-3. The tricarboxylic acid cycle

pyruvic acid can enter the tricarboxylic acid cycle, it must be converted to a more reactive form. This is accomplished by reaction of the acid with a complex molecule known as *coenzyme A*. The reactive portion of coenzyme A involves a mercaptan group (—SH), so coenzyme A frequently is written as CoASH. The reaction with pyruvic acid involves loss of carbon dioxide to form acetyl coenzyme A.

$$CH_3C-C \overset{O}{\underset{OH}{\big\|}} + CoASH \longrightarrow CH_3C \overset{O}{\underset{SCoA}{}} + CO_2$$

Pyruvic acid   Coenzyme A   Acetyl coenzyme A

Acetyl coenzyme A is an important molecule in living systems. In addition to its role in the tricarboxylic acid cycle, it is also the main building block for **fatty acids**, long-chain carboxylic acids from which fats are derived. The formation of fats provides an alternate metabolic route for excess acetyl coenzyme A molecules when the glucose concentration is extremely high. Thus, fats serve as energy-storage systems.

lipids

Lipids are naturally occurring organic compounds that are insoluble in water but soluble in organic solvents such as benzene and chloroform. They serve various functions in plants and animals, including acting as a metabolic fuel-storage system, structural components for cell membranes and cell walls, and agents for some important biological activities. Types of compounds that compose the lipid family include the fats, steroids such as cholesterol and bile acids, and the fat-soluble vitamins like vitamins A, D, and K.

Fats are glycerol esters of fatty acids. Glycerol is a trihydroxy alcohol, and each glycerol molecule can form ester groups with three fatty acid molecules. These substances serve as fuel-storage components in cells. Typical fatty acids in fats contain 16 or 18 carbon atoms in a linear chain. A typical fat is glycerol stearate, which is made from stearic acid ($C_{17}H_{35}COOH$).

The three fatty acid components of a fat may all be identical, as in glycerol stearate, or they may all be different.

Fatty acids are synthesized in the living cell by combining the acetyl groups of several acetyl coenzyme A molecules. They are decomposed by breaking off acetyl coenzyme A groups from the chain. The degradation of a fatty acid is known as β oxidation

$$
\begin{array}{l}
CH_2-O-\overset{O}{\overset{\|}{C}}-C_{17}H_{35} \\[1em]
CH-O-\overset{O}{\overset{\|}{C}}-C_{17}H_{35} \\[1em]
CH_2-O-\overset{O}{\overset{\|}{C}}-C_{17}H_{35}
\end{array}
$$

Glycerol stearate

Gallstones, which can cause severe discomfort to man, contain cholesterol, a lipid compound. (*From Waltman Walters and Albert M. Snell, Diseases of the Gallbladder and Bile Duct. W. B. Saunders and Company, Philadelphia, 1940*)

Figure 13-4. Beta oxidation of stearic acid

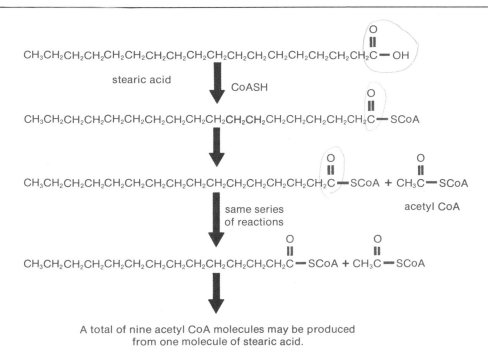

A total of nine acetyl CoA molecules may be produced from one molecule of stearic acid.

illustrated in abbreviated form in Figure 13-4. This process provides an alternate supply of energy other than glycolysis for animals. The acetyl units derived from the fatty acid degradation are used in the tricarboxylic acid cycle and lead to the production of ATP molecules. Beta oxidation occurs readily only for linear fatty acids. Branched-chain acids are degraded slowly, and then by a different process.

proteins

Proteins, the most abundant organic components in cells, are complex molecules found in all living systems. They participate in such diverse cell activities as metabolic regulation (as enzymes) and structure formation (as cell-wall constituents). Proteins are polymers that range in molecular weight from about 5000 to several million.

The most abundant protein in animals is collagen. In the human body, it constitutes approximately one-third of all protein matter. It is a fibrous material found in tendons, ligaments, cartilage, bone, skin, and other connective tissue. The basic collagen unit consists of three polymeric amino acid chains that are twisted together like a rope. Each chain contains about 1000 amino acid units and has a molecular weight of about 95,000.

The monomer units in proteins are known as **amino acids**, because they contain both an amino group ($-NH_2$) and a carboxylic acid group ($-CO_2H$).

Although many amino acids can be prepared in the laboratory, only about 20 normally occur in proteins. The structures and names of these particular amino acids are given in Figure 13-5. The human body can synthesize most amino acids from other food materials. A few amino acids, however, cannot be synthesized by the body and are thus required in the food supply. These required amino acids are called *essential amino acids* and are indicated by stars in Figure 13-5.

All naturally occurring amino acids are optically active, except glycine. It has been discovered, however, that only one type of optical isomer of the amino acids, called the *L-isomer*, occurs in nature. (See Chapter 12 for a discussion of optical activity.) The polymer chain in a protein is built up through a series of amide linkages, formed by reacting the carboxylic acid group of one amino acid molecule with the amino group of another amino acid molecule. The amide bonds in a protein are known as **peptide linkages**.

Different sequences of amino acids may be arranged along the

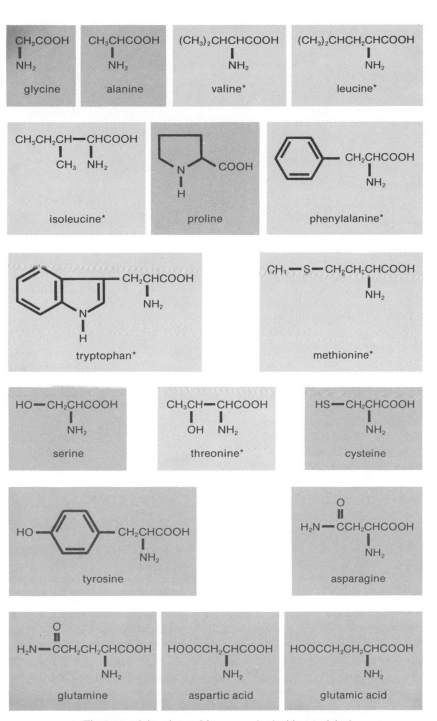

Figure 13-5. The naturally occurring amino acids

(The essential amino acids are marked with asterisks.)

Figure 13-5 (*continued*).

$$H_2N - CH_2CH_2CH_2CH_2CHCOOH$$
$$|$$
$$NH_2$$

lysine*

$$H_2N - C - NH - CH_2CH_2CH_2CHCOOH$$
$$||$$
$$NH \qquad\qquad NH_2$$

arginine

$$- CH_2CHCOOH$$
$$|$$
$$NH_2$$

histidine

$$CH_3 - CH - C \overset{O}{\underset{OH}{\big/}} \quad + H_2N - \overset{CH_3}{\underset{}{CH}} - C \overset{O}{\underset{OH}{\big/}}$$
$$|$$
$$NH_2$$

$$\searrow H_2O$$

$$CH_3 - CH - C - NH - CH - C - OH$$
$$| \qquad\quad || \qquad\quad | \quad\; ||$$
$$NH_2 \qquad\; O \qquad CH_3 \; O$$

A peptide linkage

chain to give different proteins. There is an almost infinite number of proteins theoretically possible from combinations of naturally occurring amino acids. The determination of the specific amino acid sequence of a protein is quite complicated, so this information is known only for a small number of proteins.

The first determination of the complete amino acid sequence in a protein was reported in 1954, by Frederick Sanger and co-workers at Cambridge University. The protein on which they performed their monumental study was insulin, the pancreatic hormone that governs sugar metabolism. The work extended over ten years, and Sanger was awarded a Nobel Prize for the achievement. The insulin molecule contains 51 amino acid units arranged in two chains joined by two sulfur-sulfur linkages.

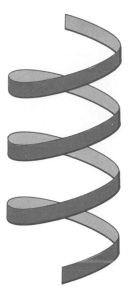
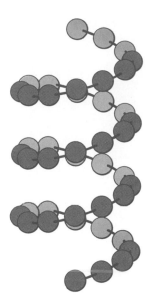

Figure 13-6. The alpha helix of proteins

The amino acid sequence in a protein is called its primary structure. Proteins also have secondary and tertiary structures that have been identified and studied. The secondary structure of proteins is related to the orientation and shape of the amino acid chain in space. Proteins have been found to exist largely in the shape of a spiral referred to as an α helix, as shown in Figure 13-6.

The protein molecule is usually so large that the alpha helix will fold up on itself to make a rather complicated structure (Figure 13-7). The folding of the molecule upon itself is referred to as the tertiary structure of the protein.

Figure 13-7. A folded protein molecule

The natural synthesis of proteins in such a manner that the proper sequences of amino acids are obtained is complex. The process has been unraveled largely within the past 15 to 20 years, principally owing to discoveries in the area of nucleic acids.

nucleic acids

Nucleic acids are polymers with molecular weights of several million. They are found in almost all living cells and viruses. The structures of nucleic acids are believed to carry the information necessary for the synthesis of proteins; nucleic acids thus regulate or determine the characteristics of the entire organism.

Electron micrograph of nucleolar genes, showing formation of ribosomal RNA precursor molecules (which appear as fibrils) in various stages of completion. (*From Miller, O. L., Jr. and Barbara R. Beatty, "Visualization of nucleolar genes," Science, 164:955–957, 1969*)

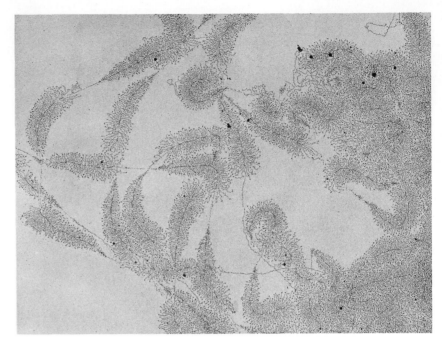

There are two basic types of nucleic acids, deoxyribonucleic acid (DNA) and ribonucleic acid (RNA). DNA is located principally in the nucleus of the cell and carries the information for the synthesis of proteins. RNA is located primarily in the cytoplasm outside the nucleus of a cell and in the ribosomes and is responsible for using the information carried by the DNA molecule.

DNA and RNA molecules are made up of a series of nucleotide molecules connected into a long chain. A nucleotide molecule consists of a five-carbon sugar molecule, a phosphate group, and an organic base molecule. The nucleotides in RNA contain the sugar ribose, and the nucleotides in DNA contain the related sugar deoxyribose.

HOCH$_2$

OH

OH   OH

Ribose

HOCH$_2$

OH

OH

Deoxyribose

Deoxyribose contains a —CH$_2$— group in place of one of the —CHOH— groups normally found in sugar molecules. In both sugars, the phosphate group is connected in the manner shown below.

$H_2O_3P—O—CH_2$ O   OH   OH   OH

$H_2O_3P—O—CH_2$ O   OH   OH

There are five organic bases that commonly occur in nucleotides. These are shown in Figure 13-8. In nucleotides, the bases are connected to the sugar molecule as shown in Figure 13-9.

In both the DNA and RNA molecules, the nucleotides are connected via the phosphate group of one nucleotide and a hydroxyl group on the sugar unit of another nucleotide.

This arrangement is known as the *primary structure* of DNA or RNA. The sequence of bases depends on the source of the DNA molecule. Thus the DNA of each organism contains a different base sequence. The significance of the base sequence will become apparent as the discussion progresses.

DNA, like proteins, exists in the shape of a helix. There is a significant difference, however, between the nucleic acid helix and that of proteins. DNA exists as a double helix; that is, there are two strands in the helix, as shown in Figure 13-10. The two strands in the helix are held together by hydrogen-bonding interaction between base groups on one strand and base groups on the other. These interactions do not occur randomly, but exist precisely between specific pairs of bases. Thus, adenine only interacts with thymine, and cyto-

Figure 13-8. Bases commonly found in nucleotides

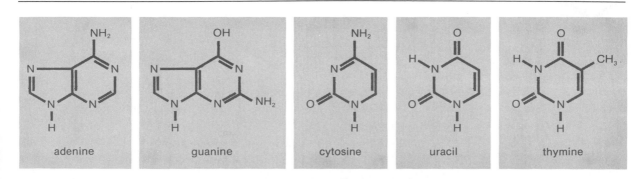

adenine   guanine   cytosine   uracil   thymine

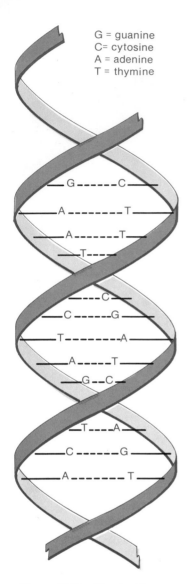

G = guanine
C = cytosine
A = adenine
T = thymine

Figure 13-10. The double-stranded helix of DNA

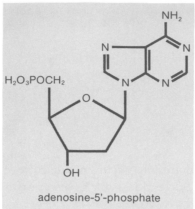

adenosine-5'-phosphate

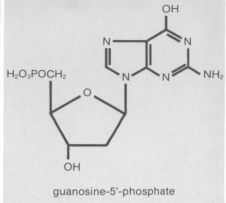

guanosine-5'-phosphate

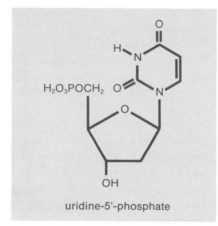

uridine-5'-phosphate

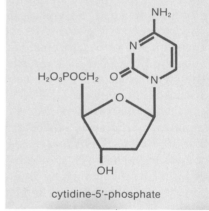

cytidine-5'-phosphate

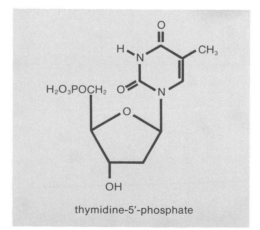

thymidine-5'-phosphate

Figure 13-9. Common nucleotides

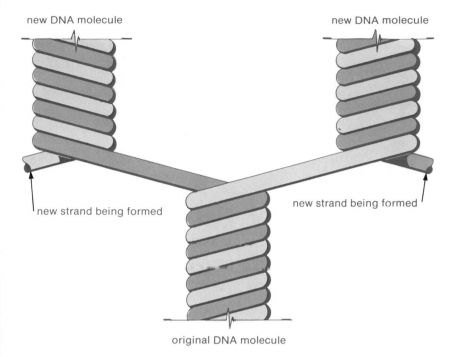

new DNA molecule new DNA molecule

new strand being formed new strand being formed

original DNA molecule

Figure 13-11. Replication of DNA

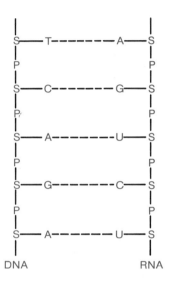

DNA RNA

Figure 13-12. Base pairing between RNA and DNA during formation of RNA

sine only interacts with guanine. Uracil is not found in DNA, only in RNA, and it interacts only with adenine. As a consequence of this specific interaction, the base sequence of one strand of DNA complements the base sequence of the second strand in the helix, as illustrated in Figure 13-10.

During cell reproduction, the two strands in the helix of each DNA molecule separate. Each strand then serves as a template for the formation of a new complementary strand. This process is shown in Figure 13-11. When the cell finally divides, one complete set of DNA molecules appears in each new cell.

In the cell, DNA serves as a template for the formation of RNA in much the same way that DNA serves as a template for its own reproduction. RNA contains uracil in place of thymine, so that the adenine-thymine combination of hydrogen-bonded base pairs in DNA is replaced by an adenine-uracil base pair when RNA is involved. The base pairing of an RNA molecule being formed on a DNA molecule is illustrated in Figure 13-12.

Since RNA is a much smaller molecule than DNA, one DNA molecule can serve as the template for many RNA molecules. Currently it is believed that the chromosomes in cells are really DNA

Human chromosomes. Chromosomes are believed to be DNA molecules that contain all the genetic information of an individual. (*Courtesy Carolina Biological Supply Company*)

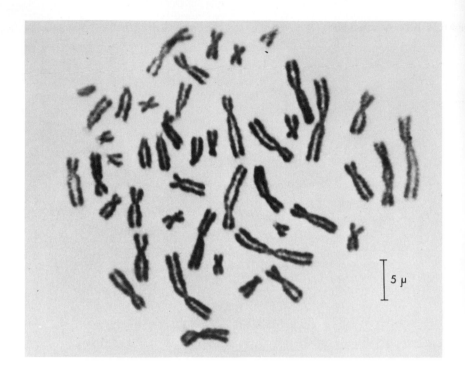

5 μ

molecules and that a gene is a small portion of a DNA molecule that is responsible for the synthesis of one RNA molecule.

There are at least two major types of RNA, known as messenger RNA and transfer RNA. In the cell, the messenger RNA attaches itself to a small globular structure known as a ribosome. The ribosome functions as the site of protein synthesis, and the messenger RNA serves as the blueprint or set of directions for carrying out the synthesis. Thus there is a unique messenger RNA molecule for each kind of protein required by the cell.

The directions for protein synthesis contained in messenger RNA are related to the sequence of bases in the molecule. Three consecutive bases along the RNA molecule constitute a "word," which is called a codon. Each codon is an instruction to the ribosome to add a particular amino acid to the emerging protein molecule. The entire set of codons for the 20 amino acids involved in protein synthesis has been pieced together by scientists and constitutes the genetic code. Thus scientists now know which amino acid corresponds to a particular codon. Some selected codons corresponding to particular amino acids are listed in Table 13-1.

Transfer RNA is a smaller molecule than messenger RNA; it has

Table 13-1.　Representative messenger RNA codons

Amino Acid Incorporated	Codon*	Amino Acid Incorporated	Codon*
Alanine	G C A	Lysine	A A A
Arginine	A G A	Leucine	C U A
Aspartic acid	G A C	Methionine	A U G
Asparagine	A A C	Phenylalanine	U U C
Cysteine	U G C	Proline	C C A
Glycine	G G A	Serine	A G C
Glutamine	C A A	Threonine	A C A
Glutamic acid	G A A	Tryptophane	U G G
Histidine	C A C	Tyrosine	U A C
Isoleucine	A U C	Valine	G U A

* G = guanine
C = cytosine
A = adenine
U = uracil

the ability to loosely bind amino acid molecules and carry them to the ribosome for incorporation into the developing protein molecules. A particular transfer RNA molecule will bind only one kind of amino acid. Thus there is at least one unique transfer RNA molecule for each amino acid required in a protein molecule. The transfer RNA molecule must have two specific binding sites. One site is for the transport of a particular amino acid, and the other is for the attachment of the transfer RNA to the appropriate codon on the messenger RNA. Once protein synthesis is complete, the messenger RNA molecule is ejected from the ribosome. It may be reinserted into a ribosome for a repetition of the protein synthesis, or it may be degraded into its individual nucleotide units and reassembled in the nucleus into a different RNA molecule.

In summary, genetic information regarding protein synthesis is contained in DNA molecules in the nucleus of a cell. Messenger RNA molecules are synthesized along a portion of a DNA molecule and carry information regarding the synthesis of a specific protein from the DNA molecule to the ribosome. In the ribosome, the messenger RNA molecule serves as a template for protein synthesis, and transfer RNA molecules carry specific amino acids to the ribosome for incorporation into a developing protein molecule. Protein synthesis is shown in Figure 13-13.

contributions and implications

Many areas of current research in the field of biochemistry can be expected to make remarkable contributions to the health and welfare of man. By investigating the biochemical origin of illnesses and the

Figure 13-13   Protein synthesis

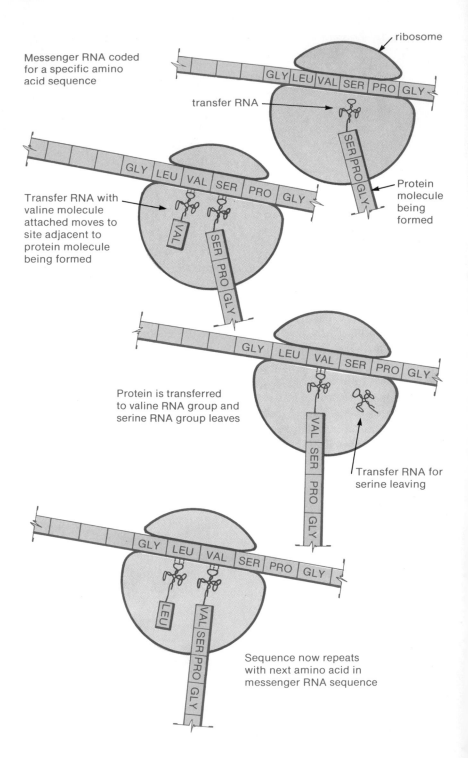

Messenger RNA coded
for a specific amino
acid sequence

ribosome

transfer RNA

Protein
molecule
being
formed

Transfer RNA with
valine molecule
attached moves to
site adjacent to
protein molecule
being formed

Protein is transferred
to valine RNA group and
serine RNA group leaves

Transfer RNA for
serine leaving

Sequence now repeats
with next amino acid in
messenger RNA sequence

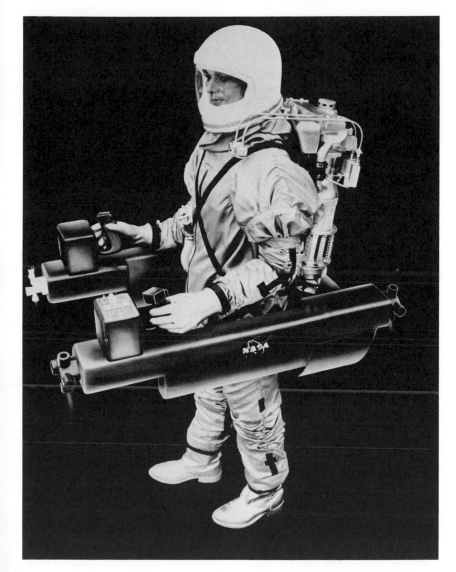

Current scientific research has many implications for the future. Perhaps someday one-man rocket-propulsion devices will be routine household and family equipment. (*Courtesy NASA*)

mechanism of drug action, scientists hope to devise new drugs with specific medicinal applications. The results might be eradication of cancer, muscular dystrophy, and other fatal or crippling diseases. It may even be possible to extend the human life span by chemically delaying the aging process. Further work in unraveling the mystery of substances that hold genetic information not only would permit repair of defective genes, thereby correcting physical and mental disorders, but also could permit synthesis of DNA molecules that would lead to creation of "synthetic" living entities.

In the early 1950s, investigations by S. L. Miller, a student of Nobel Prize winner Harold Urey, led to the formation of simple amino acids and organic acids under conditions that were believed to prevail early in the earth's geological history. Miller exposed a mixture of water vapor, methane, ammonia, and hydrogen to electrical discharges. At the end of a week, he found a mixture of simple organic molecules in his reaction vessel, including glycine, alanine, formic acid, and acetic acid. This supported the suggestion made in the late 1920s by A. I. Oparin, a Soviet biochemist, that the primitive earth's atmosphere contained methane, ammonia, and water vapor. It was postulated that these gases led to the spontaneous formation of simple organic molecules by energy from the sun or lightning discharges and that these compounds dissolved in and enriched the primitive ocean from which the first living cell eventually arose.

The potential for good in the study of biochemistry is almost limitless. At the same time, the potential for harm is just as great. Both aspects have been explored extensively by science fiction writers. Although many of these ideas were only the products of vivid imaginations several years ago, today they are within the realm of possibility. The time is not far off when society must decide what it wants and what it is willing to accept from biochemical and other scientific studies. For example, it may be possible in the not-too-distant future to create "worker" and "master" classes of people via chemical means, as proposed in Aldous Huxley's *Brave New World*. Is this the type of society that we want? Biochemists might develop a new "super" man that is stronger and more intelligent than any existing man. Again, we must ask whether society wants this.

references

Conn, Eric E., and P. K. Stumpf, *Outlines of Biochemistry*, Third Edition, John Wiley & Sons, Inc., New York, 1972.

Lehninger, Albert L., *Biochemistry*, Worth Publishers, Inc., New York, 1970.

McGilvery, R. W., *Biochemistry — A Functional Approach*, W. B. Saunders Company, Philadelphia, 1970.

study questions

1 What are the principal types of macromolecules studied in biochemistry? Which is the most abundant in cells?
2 Is glucose an aldose or a ketose? Why? What is fructose? Why?
3 What is the difference between glycogen and starch?
4 What is the difference between glycogen and cellulose?
5 What do glycogen, starch, and cellulose all have in common in terms of their chemical makeup?
6 What are the makeup and function of an enzyme?
7 By what process do plants synthesize glucose? What is the energy source for this process?
8 What is the glycolysis scheme?
9 If an amino acid is labeled as an essential amino acid, what does this mean?
10 Discuss the difference between the primary, secondary, and tertiary structures of proteins.
11 What are the structural differences between DNA and RNA?
12 Outline briefly the way in which DNA functions as the carrier of genetic information.
13 How does messenger RNA differ from transfer RNA in biological function?
14 In what part of the cell does protein synthesis occur?

special problems

1 Suppose that the nucleotide sequence in a messenger RNA molecule were as follows.

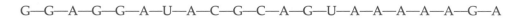

G—G—A—G—G—A—U—A—C—G—C—A—G—U—A—A—A—A—A—G—A

What would the amino acid sequence be in the protein that this RNA molecule would synthesize?
2 The product of the β oxidation of fatty acids is acetyl coenzyme A. What use might the cell make of acetyl coenzyme A?
3 Do you think that society should determine or dictate in what areas scientists may do basic research? Should society permit basic research in all areas, but restrict applied research to only certain areas?

chapter **14**

medicinal chemistry

Chemicals used to treat diseases of living organisms are known as **drugs**, although, to many people, the term refers to substances that are abused as well as to useful medicinal compounds. The drug industry is very large. It is almost impossible to watch television, listen to the radio, or read a magazine without coming across some reference to drugs. In this chapter, the chemistry and pharmacology of a few selected drugs will be discussed. Both legal and illegal drugs will be considered.

drug
pharmacology
analgesic
antipyretic
antibiotic
bacteriostatic
bactericidal
antimetabolite
steroid
androgen
estrogen
progestin
tranquilizer
alkaloid
opiate

history of pharmacology

Pharmacology is the study of the properties of drugs and their effects on living systems. In earlier days, natural herbs and other materials were used to cure disease or relieve pain. During this period, little was known about the pharmacology of these materials. It was not until the 1700s that the scientific approach to medicinal chemistry began. The advent of synthetic organic chemistry has led to synthesis of newer drugs that are often superior to naturally occurring materials.

Modern biochemical techniques have made it possible to learn how some drugs actually function in curing disease. Often these modes involve interaction of the drug with some essential metabolic process. In these interactions, the drug may interfere with the synthesis of some necessary biochemical substance such as a protein, serve as a substitute for a deficient natural substance, or modify the activity of an enzyme or a hormone.

Investigations also have led to the observation that specific physiological actions appear to be related to certain chemical structures. Likewise, compounds with similar medicinal properties often have similar basic structures, for example, barbiturates (sleeping

Useful medicinal drugs come in a huge variety of sizes, shapes, colors, textures, and tastes. *(Photograph by A. Marshall Licht)*

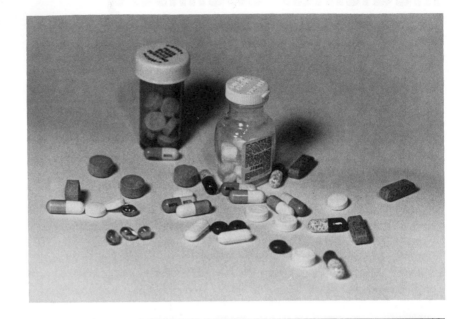

Commercial preparation of medicinal compounds. *Left*: Machine for pressing out uniformly shaped tablets. *Right*: Machine for filling capsules. *(Courtesy Parke, Davis & Company)*

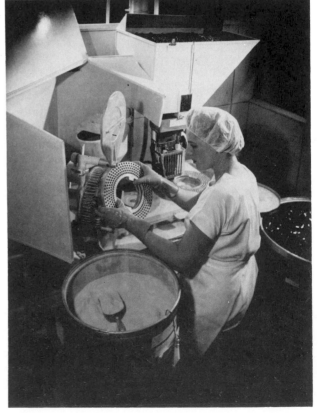

Use of medicinal materials dates to ancient man who, by instinct and by trial and error, learned what would soothe his aches and pains. Although crude, these materials and methods were the precursors of many of today's modern medicines.

During the Middle Ages, monks preserved pharmaceutical and medical knowledge by translating or copying manuscripts for their libraries and practicing the art of the apothecary by using healing herbs collected in fields or raised in monastery gardens.

Some cultures have maintained healing practices based on long-tested natural medicines and religious beliefs. Shown here is a Navajo sand-painting ceremony involving songs, prayers, and herbs shared by patient, medicine man, and spectators. Many such medicinal preparations and methods have led to the discovery of new drugs or techniques. *(Courtesy Parke, Davis & Company © 1951–1953)*

pills), amphetamines (antidepressant compounds), and penicillins (antibiotics). This information has made it possible to predict structures for new compounds that might serve as drugs for specific purposes. It also has advanced the knowledge of the nature of the specific site where a drug is believed to bring about its physiological action.

Unfortunately, the fields of biochemistry and pharmacology have not progressed to the point that drugs can be made with complete safety. Thus, testing with animals and finely controlled testing with human beings is still necessary to determine side effects and safe dosage levels for a newly developed drug. Variability in drug tolerance among individuals complicates the testing problems. In spite of careful checking, an occasional drug does reach the market that later proves to be harmful. The drug *thalidomide*, for example, showed no adverse side effects when tested on laboratory animals. Even tests on human beings showed a remarkable safety margin. The drug was introduced in West Germany in 1958, as a sedative in sleeping pills. It was not until several years later that thalidomide was found to be responsible for infants being born with deformed limbs. In the United States, the Food and Drug Administration (FDA) oversees testing new drugs and removing harmful drugs from the market.

metabolism of drugs

When drugs enter a living organism, they are subject to the same chemical reactions as other cell constituents. These chemical reactions convert the drug into new materials. In most cases, these new materials are harmless and eventually are excreted from the system. In some cases, however, the metabolized drug may be the active component in the organism; in other cases it may provoke the side reactions that accompany use of the drug. A few drugs and the metabolic changes that they undergo are shown in Figure 14-1.

legal drugs

Drugs are divided into families on the basis of principal medicinal activity. Several structural classes of compounds generally are included within each family. Drugs that generally are accepted as safe by the FDA and that do not need supervision in their use are available as over-the-counter preparations. Prescription drugs, on the

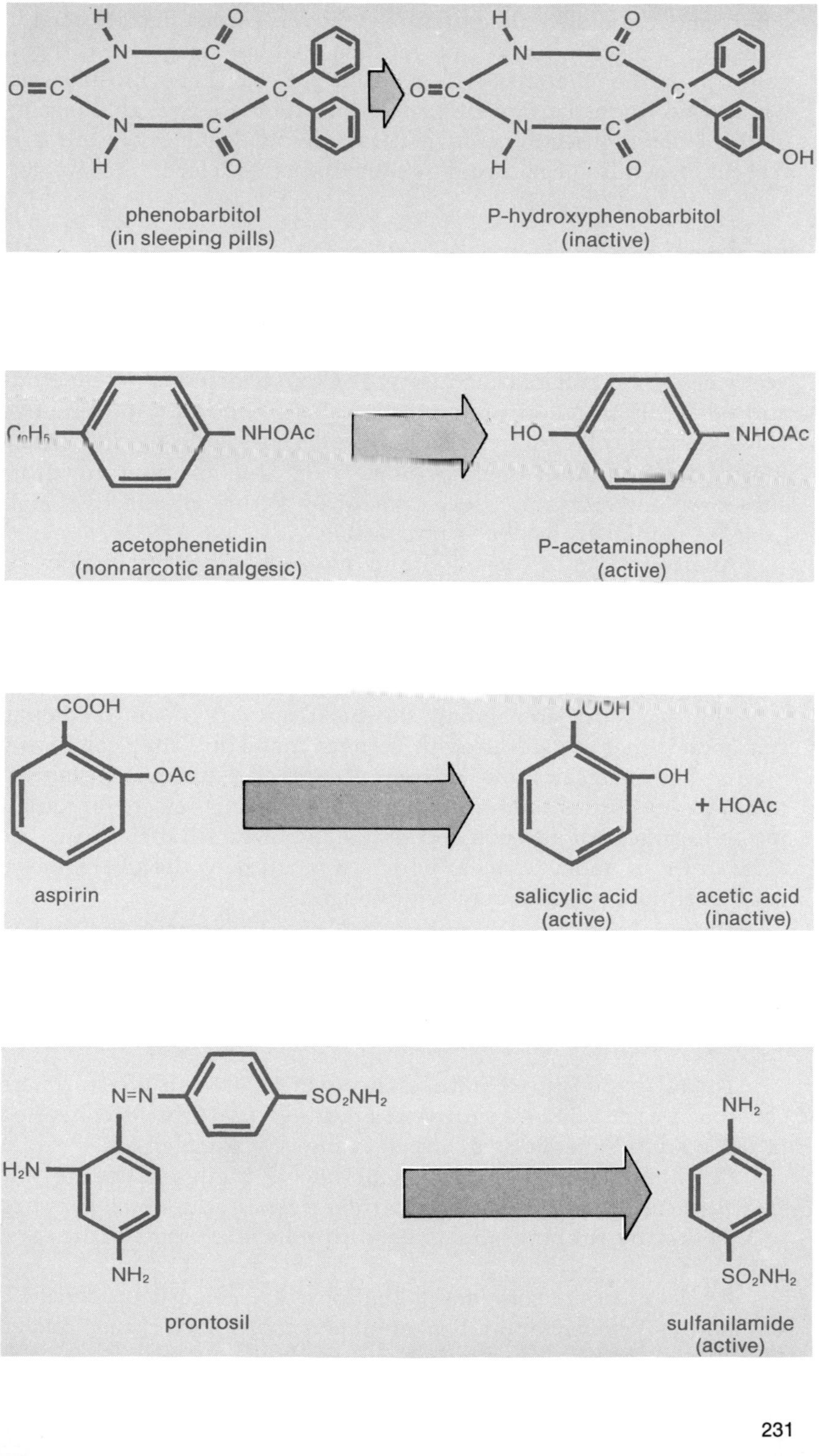
phenobarbitol
(in sleeping pills)
P-hydroxyphenobarbitol
(inactive)
NHOAc
HO
NHOAc
acetophenetidin
(nonnarcotic analgesic)
P-acetaminophenol
(active)
COOH
OAc
COOH
OH
+ HOAc
aspirin
salicylic acid
(active)
acetic acid
(inactive)
N=N
SO_2NH_2
H_2N
NH_2
NH_2
SO_2NH_2
prontosil
sulfanilamide
(active)

Figure 14-1. Metabolism of several drugs in living organisms other hand, often have the potential of adverse side effects if used in excess or indiscriminately. Nevertheless, these drugs have therapeutic properties that make them useful in treating specific illnesses, when used under supervision; therefore they are available legally. In the following sections, several families of legal drugs are discussed. Presently illegal drugs will be discussed later in the chapter.

analgesics

Relief of pain has long been a concern of man, and the search for and development of drugs for this purpose has been one of the more active areas of medicinal chemistry. The causes of pain are numerous and not easily traced or understood. Relief from pain can come two ways: removing the source of pain, or desensitizing the individual toward the sensation of pain. Since pain and its relief are quite subjective and personal, development of a truly all-effective **analgesic**, or painkiller, has been impossible.

Analgesics can be classified into two groups: narcotic analgesics, principally the opium constituents and their synthetic relatives, and nonnarcotic analgesics, such as the salicylates and the aniline derivatives. The opium compounds will be discussed in a later section of this chapter.

The *salicylates* are among the most ancient of the medicinal compounds that are still used in modern medicine. They commonly are used as analgesics and **antipyretics** (fever-reducing substances) to relieve headaches and minor pains. In addition, they are important in the treatment of arthritis and rheumatic fever. All these drugs are derivatives of salicylic acid, which was initially discovered as a constituent of a substance in willow bark.

Sodium salicylate, the sodium salt of salicylic acid, was one of the earliest members of this family to be used medicinally.

In 1853, acetylsalicylic acid (aspirin) was first synthesized, but its value as a medicinal compound was not discovered until the late 1890s.

Today, in the United States alone, over 12,000 tons of aspirin are consumed each year! As shown in Figure 14-1, aspirin is converted in the body to salicylic acid, which is the active analgesic.

The third commonly used salicylate is methyl salicylate, the principal constituent in the oil of wintergreen. It is generally applied to the skin by rubbing and is used to relieve muscular aches and pains.

Another class of compounds that possess analgesic properties is the family of *aniline* (aminobenzene) *derivatives*. Two closely related

Salicylic acid

Acetylsalicylic acid (aspirin)

compounds in particular, acetaminophen and acetophenetidine (phenacetin), have gained popular use; individuals who are allergic to or are unaided by aspirin often find relief with one of these aniline derivatives. Excedrin is a commercial product that contains these compounds as active ingredients.

The antipyretic or fever-reducing action of these drugs appears to be accomplished through a center in the hypothalamus of the brain, which causes small blood vessels in the skin to dilate, thus enabling the body to lose heat. The analgesic action of these drugs also appears to be brought about through a control center in the hypothalamus, close to the antipyretic center, but the exact mode of painkilling is unclear.

Phenacetin
(acetophenetidine)

antibiotics

Compounds that can inhibit and destroy bacteria are called antibiotics. Such compounds can be bacteriostatic (growth-inhibiting) or bactericidal (killing) in their action. Most currently used antibiotics are derived from microorganisms themselves and are in effect metabolic products of these organisms. Common classes of antibiotics include the *penicillins* and the *tetracyclines*.

Although the general concept of antibacterial compounds was known in the 1870s, the modern era of antibiotics really did not begin until 1932, with the discovery of the bacteriostatic properties of a compound called *prontosil*. It soon was deduced that the bacteriostatic properties of prontosil were due to its metabolic product, sulfanilamide, and the first family of modern, medically effective antibacterial agents, the sulfa drugs, was discovered. Sulfanilamide is believed to interfere with synthesis of folic acid, an important growth factor in bacteria. The sulfa drugs have largely been displaced in use by more recently developed antibiotics, but around World War II, sulfa drugs were the most commonly used antibacterial agents. Structures of prontosil and sulfanilamide were given in Figure 14-1.

In 1928, Alexander Fleming made his now-famous chance discovery of penicillin, but it was not until 1940 that two other British chemists, Howard Florey and E. B. Chain, isolated the antibiotic itself from the *Penicillium* mold. Since then, many penicillin derivatives have been isolated and studied, and about a dozen of these constitute the common, commercially used members of this family. The penicillins are bactericidal in their action, destroying bacteria by interfering with normal cell-wall synthesis.

Occasionally, major scientific discoveries result from unplanned, accidental incidents. This kind of situation resulted in the discovery of penicillin by Alexander Fleming in 1928. Fleming, a professor of bacteriology at the University of London, became interested in finding medicinal agents that would prevent and cure infections. While studying the common staphylococcus germ, he accidentally left one of his cultures uncovered, and it became contaminated with a spore from the mold that grows on moist bread and cheese. Fleming's observation that the microorganisms around the spore were all killed ultimately led him to the discovery of penicillin. In 1945, he shared the Nobel Prize in medicine with Howard Florey and E. B. Chain, who were responsible for the isolation of penicillin itself. In speaking of his discovery, Fleming said, "There are thousands of different molds and there are thousands of different bacteria, and that chance put that mold in the right spot at the right time was like winning the Irish Sweep."

Penicillin G, the most effective and
least toxic penicillin derivative

In 1948, aureomycin, the first member of the tetracycline family of antibiotics, was isolated from *Streptomyces aureofaciens*. All tetracyclines have been isolated from *Streptomyces* strains, and all have a basic structure of four fused rings. They are called *broad-spectrum* antibiotics because of their effectiveness against a wide variety of microorganisms.

Aureomycin

Tetracyclines have the chemical ability to bind metal ions that are essential to synthesis of proteins in bacteria, leading to the destruction of these organisms. Besides aureomycin, the tetracycline family includes terramycin, declomycin, and tetracycline itself.

anticancer drugs

A very active area of research at present is the search for drugs that will interfere with growth of cancer cells. Extensive government support for such research has been available in recent years. Because cancer is not a single specific disease, its causes are understandably diverse and not well understood. Consequently, research toward the discovery of a general cure for cancer has not been possible, although significant accomplishments in developing chemotherapeutic agents effective against certain forms of cancer have been made. In most of these cases, the therapeutic action is not specific for cancer cells, and normal cells also are affected by the potent drugs used. However, since the growth and metabolic rates of cancer cells are greater than those of normal cells, the effect of an anticancer drug is more pronounced on the cancer cells than on normal cells.

A number of the anticancer drugs currently used are called **antimetabolites**, because they interfere with the normal function of certain physiologically important substances (cellular metabolites). In most cases, these antimetabolites are close structural analogs of the physiologically important compounds. Examples of these drugs include: aminopterin and methotrexate, analogs of folic acid (an essential growth factor); 6-mercaptopurine, an analog of the purine adenine (an important constituent in nucleic acids); and 5-fluorouracil, an analog of uracil (a pyrimidine constituent of RNA). 5-Fluorouracil has been effective in inhibiting growth of certain human tumors; the other drugs have found principal use in treating acute leukemia in children.

Leukemia is a type of cancer that disturbs the normal formation of red blood cells and causes an overabundance of white cells in the blood. The antileukemia properties of mustard gas ($S(CH_2CH_2Cl)_2$) used in World War I were discovered in World War II, when soldiers who accidentally became exposed to the substance later exhibited reduced white cell counts in their blood. Because of the high toxicity and low water solubility of mustard gas, it could not be used as an effective antileukemia drug, and investigations of the related nitrogen mustards were undertaken. Of the several hundreds of compounds of the nitrogen mustard class that have been synthesized and tested, a few have been found to possess antileukemia and tumor-retarding properties. Although these compounds are less toxic than sulfur mustard gas, significant, undesirable side effects, such as nausea, local blistering, and increased chance of infection, still preclude their widespread use. One of the more commonly employed nitrogen mustards is Mustargen ($CH_3N(CH_2CH_2Cl)_2$). Its similarity in structure to sulfur mustard gas can be noted readily.

Research in the area of anticancer drugs is, and must be, aimed toward developing substances that are more specific for cancer cells and less toxic to normal cells. Simultaneous research in understanding the causes and characteristics of the various cancers is imperative if this major disease is to be conquered.

steroidal hormones and related compounds

Steroids are compounds that have the four-ring system shown below as a basic structural unit.

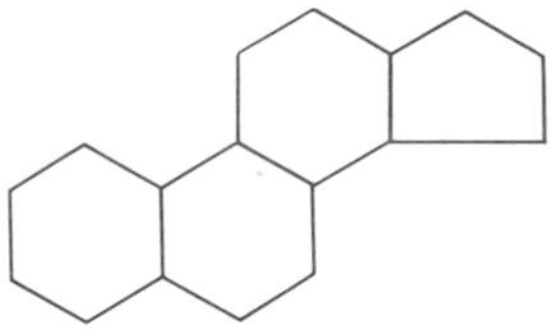

Steroid nucleus

Of the several important adrenocortical hormones, cortisone is perhaps the most familiar. This compound and its derivatives have been used in the treatment of rheumatoid arthritis, rheumatic fever, severe allergic reactions, and as anti-inflammatory agents.

The hormones derived from the sex glands are the androgens (the masculinizing sex hormones such as testosterone) and the estrogens and progestins (the feminizing sex hormones such as estradiol and progesterone). These compounds have several therapeutic uses, including removal of deficiencies of natural hormones and their consequent disorders.

Although administration of both estrogens and progestins suppresses ovulation in the female and thus prevents conception, the current oral contraceptive drugs on the market are principally progestins, with an added estrogenic compound. These include norethynodrel and norethindrone. Some commercial preparations also contain amounts of mestranol, an estrogen that prevents spotting or bleeding while the medication is used. When taken as directed, usually for 20 days of each normal 28-day menstrual cycle, these drugs have been practically 100% effective in preventing pregnancies. Certain adverse side effects such as formation of blood clots and loss of vision have been reported. These problems, and possible long-term effects on vital organs, are being investigated.

Since the effect of oral contraceptives is inhibitory on ovulation, discontinuing use of these drugs permits return of normal fertility, usually within one to two months.

tranquilizers

The name tranquilizer applies to a large, important, and diverse group of drugs. By definition, these compounds are depressants of the central nervous system, with highly selective action on brain function. They differ, however, in the type and level of psychophar-macological effect. In proper doses these drugs generally can calm an individual without dulling mental alertness or causing sleep, as observed with sedatives such as the barbiturates.

Tranquilizers have been divided arbitrarily into major and minor classes, based on the level and quality of calmness and relaxation induced. Major tranquilizers, which are used to treat psychoses, include the phenothiazines, such as chlorpromazine (Thorazine), the first effective member of this family of drugs, and some *Rauwolfia* alkaloids, most notably reserpine (Serpasil) Minor tranquilizers, which are used to relieve tension and anxiety, include meprobamate (Miltown, Equanil) and chlordiazepoxide (Librium).

Reserpine is a compound found in the roots of several *Rauwolfia* species.

Reserpine

It is an alkaloid, which is a class of naturally occurring, physiologically active, nitrogen-containing compounds of plant origin. Preparations of *Rauwolfia* alkaloids were used in India for many years as a treatment for various ailments, including insanity, but it was not until 1952 that reserpine was isolated from these plants. The value of this alkaloid as a tranquilizer was recognized shortly thereafter. Reserpine lowers the level of serotonin, an important brain amine; this procedure has been suggested as a possible cause for the drug's tranquilizing effect. In addition, the drug acts as an antihypertension agent in reducing high blood pressure.

barbiturates

Barbituric acid

Drugs that induce relaxation and sleep often are needed to treat patients with organic and emotional disorders. Several families of drugs are used for this sedative and hypnotic purpose, but the most common and most widely used is the *barbiturate* family. Drugs of this group are derivatives of barbituric acid, with various substituents principally at position 5.

The first clinically used barbiturate was introduced in 1903 in Germany. This compound was barbital, known commercially as Veronal. In 1912, the second barbiturate of wide acclaim, phenobarbital (Luminal), was introduced. Although more than 2000 barbiturate derivatives have been synthesized since then, only about 25 have commonly been used medicinally. Of major concern in selecting a barbiturate for sleep induction or sedation are the desired time of onset of action and the desired duration of effects. Certain barbiturates, particularly the sulfur-containing ones such as sodium pentothal, have been used successfully as intravenous anesthetics because of their rather short duration of activity.

Barbiturates have been referred to as one of the most useful families of drugs, because of their range of medicinal effects. These effects are determined by the doses in which the drugs are administered. In small doses, barbiturates act as sedatives. In doses three to five times sedative doses, these drugs act as sleep-inducing compounds. In still larger doses, they serve as general anesthetics. Overdoses cause death.

The exact mechanism of action of the barbiturates has not been determined conclusively yet. There seems to be general agreement, however, that some depression of the central nervous system is involved.

Like nearly all drugs, barbiturates are generally quite safe when used in properly prescribed dosages. They are habit-forming, however, and overdoses due to ingestion of large numbers of sleeping pills containing barbiturates are not uncommon. In the United States, about 15% of all poisonings and deaths from poisons are due to barbiturates. Most stem from attempted suicides, but since large doses cause intoxication similar to alcoholic intoxication, sleeping pills or "goof balls" have been used illegally for "kicks," sometimes with dangerous results.

antidepressant drugs

Antidepressant drugs are used to treat depression disorders, one of the conditions associated with certain forms of mental illness. These drugs affect the central nervous system and are used under medical direction to bring about an increase in alertness, wakefulness, and attentiveness, as well as an elevation of mood in depressed persons. Unfortunately, the use of a number of these drugs has been abused. Nonprescribed use is prohibited because of dangerous side effects associated with overdoses of these drugs.

Until the 1950s, amphetamines were used widely as antidepressant drugs, because of their relatively mild toxicity and because they were about the only effective drugs available for this purpose. Amphetamine itself is optically active. It is available as a racemic mixture (Benzedrine) and as one of the pure optical isomers (Dexedrine). Methamphetamine, a methylated derivative of amphetamine, is also commonly used. A side effect of amphetamines is loss of appetite, which has led to occasional use of these drugs in weight control.

$$\text{Ar}-CH_2-CH-NH_2$$
$$\underset{\displaystyle CH_3}{|}$$

Amphetamine

The search for better antidepressants led to the discovery of iproniazid (Marsilid). Iproniazid was developed in 1951, originally as an antituberculosis drug, but later was discovered to be more valuable as a drug for helping depressed and withdrawn mentally ill patients. Since discovery of iproniazid, other, more effective drugs of this class have been synthesized, including Niamid, Marplan, and Nardil (phenelzine). All these compounds have the physiological property of inhibiting the enzyme monoamine oxidase, which is responsible for the metabolic degradation of certain highly potent chemical compounds in the brain and nervous tissue.

illegal drugs

Several types of drugs are not used extensively, if at all, in medicine because of high toxicity or serious adverse effects. Typical dangerous effects are a tendency for addiction, uncontrolled behavioral changes, and long-term permanent damage to tissues and organs. Some well-known, currently illegal drugs are discussed in the next few sections.

opiates

Opium, the dried sap of the unripe fruit of the poppy (*Papaver somniferum*), has been used for almost 6000 years. Until the last 150 years or so, it was the only drug that doctors could rely on to relieve pain, induce sleep, and relieve dysentery. The nonmedicinal use of opium to create a feeling of well-being or to relieve anxiety probably has been known for as long as its medicinal use. References to both medicinal and nonmedicinal uses of opium can be found throughout history. The British even were involved in wars for the right to provide illegal opium to the Chinese during the nineteenth century.

In 1806, Friedrich Serturner reported the isolation of the major active ingredient of opium, now called *morphine*. This discovery opened the whole field of alkaloid chemistry. Another active component of opium, *codeine*, was isolated in 1832. The medicinal uses of morphine and codeine soon became obvious, and they quickly were adopted as the preferred painkillers of the day. The extensive use of morphine during the American Civil War left many soldiers seriously addicted. Furthermore, opium derivatives commonly were used in the patent medicines popular in the latter half of the nineteenth century, leading to the addiction of many housewives. It has been estimated that by the end of the nineteenth century, 1% of the United States population may have been addicted to opiates (drugs derived from opium) in one form or another. Opiates finally were banned from the nonprescription drug market in 1914, by the Harrison Act.

In 1874, it was discovered that morphine could be modified chemically by attaching two acetyl groups to the molecule. The resulting compound, now called *heroin*, was marketed in 1898 as a nonaddicting substitute for morphine and codeine. The premise that heroin was nonaddictive turned out to be false. The effect of heroin on the body is identical to that of morphine, except that heroin is about three times more potent. The reason for the increased potency is that the acetyl groups on the molecule make heroin more soluble in the fatty tissues of the body, and thus it enters the brain more rapidly than morphine. Once in the cell, the acetyl groups are removed by hydrolysis to form morphine.

As a consequence of the 1914 Harrison Act and later Supreme Court decisions, opium became almost impossible to obtain except on the illegal market. Nonmedical possession and smoking of opium became illegal, and thus opiate users became criminals. By World War II, the supply of opiates in the United States had almost been eliminated, and the number of users was relatively small.

Morphine

Heroin

More recently, the trend has begun to reverse itself. The number of users has increased, and the average age of addicted individuals has dropped. In 1960, less than 10% of the deaths caused by heroin were of teenagers. By 1969, 30% of those who died from heroin use were teenagers. Heroin is now the leading cause of death among teenagers in New York City.

The type of individual most commonly addicted to opiates also has changed over the years. In the nineteenth century, the common addict was a Civil War veteran or a middle-class housewife. After World War II, the typical addict was a black city-dweller in the lower income brackets. More recently, middle-class youth has been most commonly associated with illegal or abused drugs. The ready availability of heroin in southeast Asia also led to the addiction of some American soldiers during the Vietnam conflict.

A common cause of death among heroin addicts is an overdose. The reason for this is probably the variation in quality and concentration of heroin sold on the illicit market. Illegal heroin normally is diluted ("cut") with either quinine or lactose (milk sugar) before it is sold. One study showed a variation from 1% to 77% in the concentration of heroin in packages sold on the street. In addition, the probability of infection while taking heroin is quite high, and the narcotic effect of the drug tends to cover up the signs of serious illness. Furthermore, addicts frequently suffer from malnutrition due to lack of interest in food or inability to afford it.

Treatment of ex-addicts has had mixed effectiveness over the

Bags containing a total of four pounds of crude opium confiscated by U.S. Customs officials. White crystalline heroin is prepared from the morphine in crude opium. *(Courtesy U.S. Customs Service, Washington, D.C.)*

In an attempt to smuggle heroin, a false bottom was constructed for this straw-covered wine jug. A partition separated a small amount of wine in the top from the jug's lower chamber, where heroin was concealed. *(Courtesy U.S. Customs Service, Washington, D.C.)*

years. As already mentioned, heroin first was introduced as a non-addictive alternative to morphine, but was a complete failure. More recently, methadone has been used with at least moderate success.

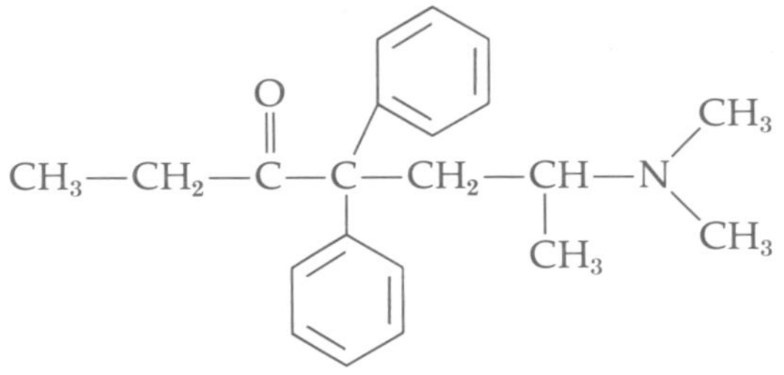

Methadone

When used in a carefully controlled program, methadone satisfies the addict's need for heroin without producing some of heroin's extreme effects. Unfortunately, the earlier bad experience with heroin has made both officials and the general public somewhat hesitant about using methadone or any substitute. More recent evidence indicates that methadone itself is also addictive.

hallucinogens

Hallucinogenic drugs have been known for thousands of years. Peyote, for example, is a cactus (*Lophophora williamsii*) found in the southwestern part of the United States. It has been used in religious ceremonies for as long as written records have been kept in the area. The principal active ingredient is mescaline, which causes euphoria, vivid colors, and other imagined visual effects.

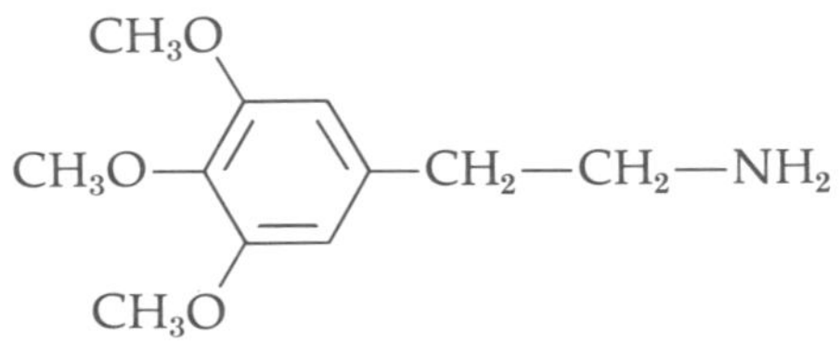

Mescaline

In 1938 a Swiss chemist, Albert Hofmann, synthesized lysergic acid diethylamide (LSD).

dl-Lysergic acid diethylamide (LSD)

The hallucinogenic properties of LSD were discovered accidentally by Albert Hofmann, a research chemist with Sandoz Chemical Works in Basel, Switzerland. Hofmann was investigating derivatives of ergot alkaloids obtained from a fungus that infects rye kernels. One of the synthetic compounds he was studying was lysergic acid diethylamide (LSD). On April 16, 1943, he somehow accidentally ingested a small amount of LSD. An entry in his research notebook describes his initial reaction.*

". . . I was seized with a feeling of great restlessness and mild dizziness. At home, I lay down and sank into a not unpleasant state of delirium, which was characterized by extremely excited fantasies. In a semiconscious state, with my eyes closed (I felt the daylight to be unpleasantly dazzling), fantastic images of extraordinary realness and with an intense kaleidoscopic play of colors assaulted me."

Hofmann traced this unusual experience to LSD and intentionally took a 250-microgram dose of the drug (far larger than the dose needed to produce hallucinations). The experiences of his second encounter with the drug were described as follows.

"As far as I can remember, the following symptoms were strongly marked during the height of the crisis, which had already passed when the doctor arrived: dizziness and defective vision; the faces of those present seemed to me like colored masks; strong kinetic disturbances alternating with paralysis; my head, my whole body and limbs seemed at times very heavy, as if filled with lead; cramps in my legs, my hands sometimes cold and numb; a metallic taste on my tongue; my throat dry and contracted; a feeling of suffocation; I was alternately bewildered and in clear understanding of the situation, so that I sometimes stood outside myself as a neutral observer as I shrieked half madly or babbled unintelligible nonsense."

*W. A. Stoll, Schweiz. Arch. Neurol. Psychiat. 60, 279 (1947); quoted in John Cashman, The LSD Story, Fawcett Publications, Inc., Greenwich, Conn., 1966.

The hallucinogenic activity of this chemical was not recognized until 1943, however, when Hofmann accidentally ingested some of the compound. The effects of even very small quantities of LSD are quite pronounced. Typical symptoms are: visual distortion of objects, changing colors, motor unrest, confusion alternating with a feeling of heightened awareness, and a feeling of having left the body. Sounds frequently have a visual effect.

The illegal manufacture and use of LSD began in the middle 1960s, peaked around 1967 or 1968, and has declined somewhat since then. The dangers of LSD have become common knowledge to most people, which probably led to its decline in use. Frightening experiences, called *bad trips*, and accidental deaths or serious injury while under the influence of the drug were commonplace. A report that LSD can cause chromosome damage, and thus mutations in children, contributed to its decline. The validity of this latter report has apparently not been fully verified, however.

marijuana

Marijuana is the common name for *Cannabis sativa*, a plant originating in the Orient and now known throughout the world. One variety of the plant is grown for its long fibers, which are used to make hemp rope, and commonly is found in the United States.

The psychoactive agent has been the subject of considerable study. Presently, there appear to be four active components. Of these compounds, Δ^9-*trans*-tetrahydrocannabinol seems to be the most active. The psychoactive agents are concentrated in the resin of the female plant. The material sold as hashish consists primarily of the resinous parts of the plant and is quite potent. The leafy material and small stems are sold as marijuana and are less potent.

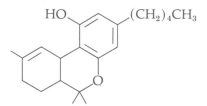

Δ^9-*trans*-Tetrahydrocannabinol

Section of leaves from a wild marijuana plant *(Cannabis sativa). (UPI Photo)*

While the short-term effects of marijuana are well known, the long-term effects still are debated. Many people claim that marijuana is no more harmful than alcohol and thus should be legalized. Others claim that it leads to the use of more potent, addicting drugs like heroin. Long-term psychological effects of the drug do not seem to have received much scientific attention so far. Until these questions are settled adequately, with good scientific data, the controversy over the legalization of marijuana probably will continue.

illegal drug use and the individual

Looked at in the most general way, nearly every adult in the United States is a drug user to some extent. Most people drink coffee, tea, or liquor or smoke cigarettes without recognizing that these contain relatively mild legal drugs. Some people depend on prescription drugs such as tranquilizers and sleeping pills. Still others use currently illegal drugs such as marijuana, LSD, and heroin. Few people avoid all these drugs, and the number of illegal drug users seems to be increasing. Some people even indulge in excessive use of such nonprescription, over-the-counter medications as cough syrups, mild tranquilizers, sleeping pills, and laxatives.

Under these conditions, it is difficult to develop a rational policy toward drug use. Abuse of certain drugs cannot be tolerated socially, since a person who abuses these drugs cannot function as a useful member of society. On the other hand, one must ask whether moderate use of presently illegal or nonprescription, over-the-counter drugs is necessarily undesirable. The existing legal position is that moderate use of caffeine, nicotine, alcohol, and over-the-counter drugs is acceptable, but use of marijuana, LSD, and heroin in any amount is unacceptable. The legal attitude, of course, reflects the attitude of society in general. Presently, however, the attitude of society toward drug use seems to be changing. As with any period of change, legal and social attitudes are not necessarily in harmony. Eventually, however, a rational approach to the problem may emerge. This approach must be based on the known short-term and long-term effects of both moderate and excessive use of drugs on the human body.

Meanwhile, what position should the individual take on the issue? The answer, of course, will vary with the individual. Moderate use of certain psychoactive drugs probably will be shown eventually to be safe for most individuals, when the drugs are prepared by reputable pharmaceutical companies and are used under the direction of trained personnel. Unfortunately, it is not known at the present time which drugs will be found safe; many already have been found to be dangerous. Furthermore, illegal sources of drugs are unreliable with respect to purity and concentration. For these reasons, it is safest to avoid the use of illegal drugs entirely and to use prescription and over-the-counter drugs only in moderation to cure or relieve specific health problems.

references

Burger, Alfred, Editor, *Medicinal Chemistry*, Second Edition, Interscience Publishers, Inc., New York, 1960.

DiPalma, Joseph R., Editor, *Drill's Pharmacology in Medicine*, Third Edition, McGraw-Hill Book Company, New York, 1965.

Goth, Andres, *Medical Pharmacology*, Sixth Edition, The C. V. Mosby Company, St. Louis, 1972.

Ray, Oakley S., *Drugs, Society, and Human Behavior*, The C. V. Mosby Company, St. Louis, 1972.

Faltermayer, Edmund K., *What We Know about Marijuana So Far*, Fortune, **83** (3), 96 (1971).

study questions

1 In what ways may a drug bring about its physiological action?
2 Is the drug that is administered always the active physiological species?
3 Distinguish between legal and illegal drugs.
4 With what medicinal property are the following families of compounds generally associated?
   a. Salicylates   d. Barbiturates
   b. Phenothiazines   e. Opiates
   c. Tetracyclines   f. Amphetamines
5 How does an antipyretic work?
6 What was the significance of prontosil in the development of antibiotics?
7 What are broad-spectrum antibiotics?
8 What are the two major effects that an antibiotic may have on bacteria?
9 What problems are associated with development of an anticancer drug?
10 To what class of compounds do oral contraceptive drugs belong?
11 What are alkaloids?

12 What is the common name given to the dried sap of the unripe fruit of the poppy? What physiologically active compounds does it contain?

13 How are morphine and heroin structurally related?

14 What is the difference between marijuana and hashish?

special problems

1 What might be the arguments for and against legalizing marijuana?

2 What testing procedures should be followed before a newly obtained compound is marketed as a drug?

3 Do you believe all advertising concerning medicinal products? What responsibilities should a drug manufacturer have in advertising medicinal products?

4 It has been suggested that many of the amphetamines, barbiturates, and other synthetic drugs used illegally originally were manufactured by legitimate pharmaceutical companies. What responsibilities should these companies have in minimizing the flow of their products into the illegal market?

chapter **15**

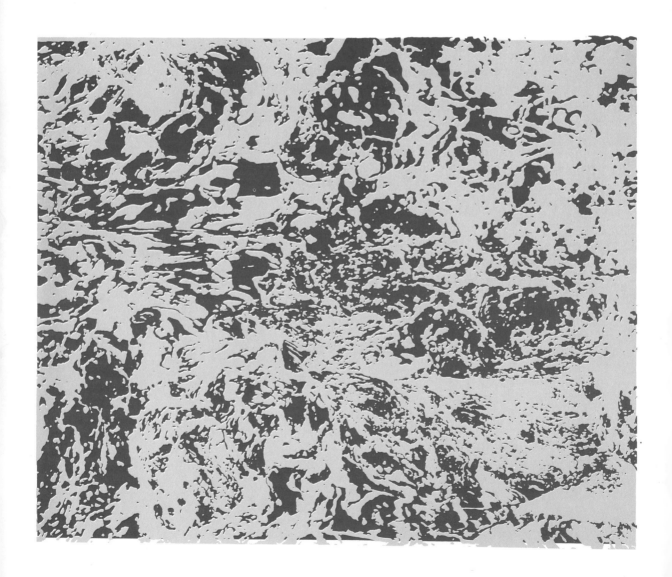

household chemistry

Although we are unaware of it most of the time, hundreds of chemical reactions take place every day in the typical home. In normal daily operations, the average homemaker cooks meals, washes clothes and dishes, and complains about soap scum. From time to time the lawn must be fertilized, the garden must be sprayed, and the house must be painted. All these activities and many more involve chemistry. Even the simple operation of heating the home normally requires a chemical reaction, either in the home itself or at an electrical power plant. Some of these applications of chemistry are discussed in this chapter; others are discussed elsewhere.

leavening
soap
detergent
surfactant
hard water
temporary hard water
permanent hard water
ion exchange
demineralized water

cooking

In the past, cooking typically has been a chore of the housewife. Increasingly, it is also becoming a hobby of men, and for a number, it has become a profession. Most people are aware of the art of cooking. They know that, among other things, cooking can improve taste, make food more tender or more digestible, and preserve food by destroying bacteria or other organisms. People also know that if they are not careful, cooking can destroy food's natural flavor or nutrients and form material that is less digestible than when uncooked. Most people know almost nothing, however, about the chemical reactions that bring about these changes.

There are a number of ways to cook food, including frying, boiling, roasting, baking, and broiling. The chemistry of two of these, frying and baking, will be considered here.

Frying can be done in either animal fat or vegetable oil. During the frying process the food is changed chemically. Proteins, for example, are coagulated, or made insoluble in water. The coagulation of proteins may be seen during the process of frying an egg.

The fats and oils used in cooking are glycerol esters of fatty acids. (See Chapter 13 for an additional discussion of fats.)

$$CH_2-O-\overset{\overset{\displaystyle O}{\|}}{C}-(CH_2)_{16}CH_3$$

$$CH-O-\overset{\overset{\displaystyle O}{\|}}{C}-(CH_2)_{16}CH_3$$

$$CH_2-O-\overset{\overset{\displaystyle O}{\|}}{C}-(CH_2)_{16}CH_3$$

Glycerol stearate
(a typical fat)

Animal fats primarily contain long-chain, saturated fatty acids such as stearic acid ($C_{17}H_{35}COOH$) and palmitic acid ($C_{15}H_{31}COOH$). Vegetable oils usually contain slightly shorter, unsaturated fatty acids such as oleic acid ($CH_3(CH_2)_7CH{=}CH(CH_2)_7COOH$), linoleic acid ($CH_3(CH_2)_4(CH{=}CHCH_2)_2(CH_2)_6COOH$), and linolenic acid ($CH_3CH_2(CH{=}CHCH_2)_3(CH_2)_6COOH$). Animal fats are greasy solids, and vegetable oils are usually viscous oils. Before vegetable oils became readily available, animal fats were the main source of cooking fats. Acceptance of vegetable oils for this purpose was slow because they did not resemble lard or other, more familiar fats. It was soon found, however, that a solid material closely resembling lard in appearance could be made from vegetable oils by catalytic hydrogenation of the double bonds. Products of this type are marketed today.

$$
\begin{array}{ccc}
CH_2OOC(CH_2)_7CH{=}CH(CH_2)_7CH_3 & & CH_2OOC(CH_2)_{16}CH_3 \\
| & & | \\
CHOOC(CH_2)_7CH{=}CH(CH_2)_7CH_3 & \xrightarrow[Ni]{H_2} & CHOOC(CH_2)_{16}CH_3 \\
| & & | \\
CH_2OOC(CH_2)_6(CH_2CH{=}CH)_2(CH_2)_4CH_3 & & CH_2OOC(CH_2)_{16}CH_3
\end{array}
$$

A typical vegetable oil

Margarine is made in the same manner, except that coloring and flavoring materials are added to make the product resemble butter.

Over the years, ordinary vegetable oils have become more popular. This popularity is due in part to recent announcements from the medical profession relating the consumption of saturated fatty acid to heart disease. Today vegetable oils are advertised as polyunsaturated, to take advantage of the medical disclosures. The term *polyunsaturated* refers to the fact that these oils contain primarily unsaturated fatty acid units.

Frying food has several disadvantages that make it undesirable as the sole method of cooking. Fats and oils usually are absorbed by food and are difficult to digest. A hard crust usually forms on the surface of fried food, and this too is hard to digest. Furthermore, usually some of the fat or oil is decomposed thermally during the frying process, and the result of this decomposition is the conversion of the glycerol of the fat or oil to acrolein. Acrolein is a noxious material that is responsible for the odors and eye irritation associated with burning fats.

$$CH_2\!\!=\!\!CHCHO$$
Acrolein

Baking is done in an oven, usually between 250°F and 450°F. Bread is an example of a food that is prepared by baking. Bread is made from flour, water, fat, sugar, and salt. Yeast is added as a leavening or "raising agent." Typical proportions of the ingredients are listed in Table 15-1.

Table 15-1. Typical proportions of bread ingredients

Ingredient	Amount (in parts by weight)
Flour	100
Water	60–63
Fat	0–3.3
Sugar	2–5
Salt	1.5–2
Yeast	2–3

The salt helps to regulate the rate of fermentation of the yeast. Without salt, fermentation is too fast, and the bread becomes too porous or "light." With too much salt, fermentation is too slow, and the bread is more compact or "heavy." The fat, or shortening, helps to prolong the bread's fresh taste.

Bread is made to rise by the formation of carbon dioxide bubbles in the dough. The process is known as leavening. The carbon dioxide bubbles come from the action of yeast on sugar. Yeast is a microorganism that can convert sugar into alcohol and carbon dioxide. (Yeast is also responsible for the fermentation of grape and other fruit juices to produce wine.)

$$C_6H_{12}O_6 \rightarrow 2\ CO_2 + 2\ C_2H_5OH$$
Sugar $\qquad\qquad\qquad$ Ethyl alcohol

Above: Processing champagne in a California winery. Yeast fermentation of grape juice takes place in bottles for over two years before the yeast is removed by sedimentation (in the large tanks at the left). The champagne is then rebottled and aged for an additional three to six months. *(Courtesy Ernest Braun, Paul Masson Vineyards)*

Above right: Illegal still taken from a New York City cellar during Prohibition. *(UPI Photo)*

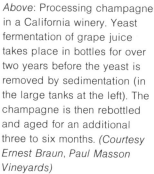

A typical baking temperature for bread is about 220°C. Baking time, and to some extent temperature, depend on the loaf's size. During baking, the dough is converted from an unpalatable, undigestible mass into a soft porous material, which is more easily penetrated by digestive juices than unbaked dough. Furthermore, much of the starch in the flour is converted to dextrin, a type of sugar, which is more easily digested than starch.

Cakes and other pastries also are made to rise (are leavened) by the action of carbon dioxide, but in these foods the source of carbon dioxide is baking powder. Baking powder usually consists of a mixture of baking soda ($NaHCO_3$) and a solid acid. Typical acids that have been used are tartaric acid ($H_2C_4H_4O_6$), cream of tartar ($KHC_4H_4O_6$), and sodium dihydrogen phosphate (NaH_2PO_4). When water is added, a chemical reaction occurs that results in the evolution of carbon dioxide.

$$KHC_4H_4O_6 \; + \; NaHCO_3 \; \xrightarrow{H_2O} \; CO_2 \; + \; H_2O \; + \; KNaC_4H_4O_6$$

| Cream of tartar (potassium hydrogen tartrate) | Baking soda (sodium bicarbonate) | Rochelle salt (sodium potassium tartrate) |

In commercial baking powder, both components are dry-mixed in advance by the manufacturer, since no reaction will take place until water is added. Moisture from the air causes baking powder to slowly decompose, however, so it must be kept tightly sealed until used.

Frequently starch is added to baking powder to retard the action of moisture, apparently by absorbing the moisture itself.

Baking powder may be made in the kitchen by mixing cream of tartar and baking soda in the proper proportions required by the equation above. In teaspoonfuls, this would be about $1\frac{1}{2}$ teaspoonfuls of cream of tartar to $\frac{1}{2}$ teaspoon of baking soda. Some older recipes call for these ingredients rather than baking powder.

soaps

Common soaps are sodium or potassium salts of fatty acids and have been used for centuries for cleaning. Until relatively recently, they were made in the home from animal fat and lye (sodium hydroxide). The animal fat was obtained as a by-product of cooking, and the lye was obtained by allowing water to leach through a barrel of wood ashes. Currently, soaps are made commercially by heating a fat with a slight excess of sodium or potassium hydroxide solution. Since animal fats contain several different fatty acid residues, soaps are actually mixtures of the sodium or potassium salts of these acids.

Tallow from cattle and sheep contains fatty acids primarily with 16 or 18 carbon atoms. Soap made from this source has good cleaning action, but is only slightly soluble in cold water. Commercial soap usually consists of salts of acids derived from nut oils, which tend to have more of the shorter acids containing 12 or 14 carbon atoms. These soaps are more water-soluble than those made from longer-chain acids. Typical nut oils are coconut oil, babbasu oil, and palm kernel oil. Castile soap, which originally was made from olive oil, now is made from cottonseed oil and peanut oil.

Soaps made by the action of sodium hydroxide on fats and oils are most common, because they can be formed into cakes or bars relatively easily. Commercial soaps frequently contain perfumes, dyes, or antiseptics. Scouring soaps contain sand, pumice, or other abrasive materials. "Floating" soaps contain air bubbles, which lower the soaps' specific gravity to about 0.8 to 0.9.

Soap made by the action of potassium hydroxide on fats and oils is soft and more water-soluble. Wood-ash lye usually contains large amounts of potassium hydroxide. Consequently, soaps made by using this lye are largely potassium soaps. Today, potassium soaps are used for special purposes such as shaving cream.

Although soaps are commonly thought of as cleaning agents, technically any salt of a fatty acid is a soap. Soaps containing metal ions other than sodium or potassium usually are not effective cleaning agents. Calcium ions, for example, make the soap insoluble in water. Aluminum soaps are commonly the major constituents in grease and are used for lubrication.

The basis for the cleansing action of soaps is that they contain a long hydrocarbon chain with an ionic group at one end.

$$CH_3CH_2CH_2CH_2CH_2CH_2CH_2CH_2CH_2CH_2CH_2CH_2CH_2CH_2CH_2C \overset{\displaystyle O}{-}O^{-1}Na^{+1}$$

Long hydrocarbon chain   Ionic group

The ionic group is water-soluble, and the hydrocarbon group is oil-soluble. Many dirt particles are oil-containing. As a consequence, the hydrocarbon end of the soap is attached to and dissolves in dirt particles, leaving the ionic end exposed to the water. The net effect is to make the dirt particle disperse in the water as a stable emulsion and thus remove it from the object to be cleaned. Figure 15-1 (on page 255) shows a dirt particle with soap molecules attached.

detergents

As will be discussed later, the cleansing action of soap is decreased by hard water. For this reason, soap has been largely replaced by synthetic **detergents**. Synthetic detergents are materials that resemble

The soap-making process. *Left*: Pioneer methods, being re-created for demonstration purposes. *Right*: A large, modern soap and detergent factory. *(Courtesy Procter & Gamble Company)*

soaps in their cleaning action, but are more resistant to precipitation by hard water. There is considerable variation in the structure and composition of detergents, but the active ingredients, or **surfactants**, all have the same principal features as soaps, and they function in essentially the same manner. Detergents, like soaps, contain a long hydrocarbon chain and an ionic group, usually a sulfonate group. Two typical surfactants are sodium salts of lauryl sulfate and alkylbenzene sulfonate.

$$C_{12}H_{25}OSO_3^{\ominus}Na^{\oplus} \qquad C_{9\text{-}15}H_{19\text{-}31}\text{---}\langle \rangle\text{---}SO_3^{\ominus}Na^{\oplus}$$

Lauryl sulfate   Alkylbenzene sulfonate (ABS)

Alkylbenzene sulfonate, or ABS, is probably the most common surfactant. It is made by condensing a linear hydrocarbon with benzene and then sulfonating the product. The linear hydrocarbon is either isolated from a cracked petroleum fraction or made by condensing ethylene. The product is a mixture of olefins with about the desired number of carbon atoms.

$$6\,CH_2\!\!=\!\!CH_2 \longrightarrow CH_3CH_2CH_2CH_2CH_2CH_2CH_2CH_2CH_2CH_2CH\!\!=\!\!CH_2$$

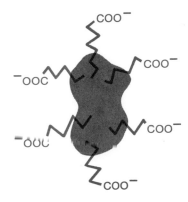

Figure 15-1. A dirt particle with soap molecules attached

In addition to the surfactant, most detergents also contain other materials known as *builders*. The most common material used as a builder is sodium polyphosphate ($Na_5P_3O_8$). Many detergents contain 35–65% polyphosphate, although a number have higher or lower percentages. To some extent, the amount of polyphosphate in a detergent depends on the detergent's intended use. Liquid dishwashing detergents tend to be lowest in polyphosphates, and laundry detergents tend to be highest. Table 15-2 lists some common detergents and their polyphosphate contents.

Notice that the percentage of phosphorus also is listed in the table. Most manufacturers list the percentage of phosphorus on the package. The percentage of polyphosphate is higher than the percentage of phosphorus because polyphosphate contains eight oxygen atoms in the molecule.

Phosphate in detergents serves several useful purposes. It functions as a water softener by tying up hard-water ions in the water. It also tends to aid the surfactant by dispersing the suspended dirt. Finally, because of its ability to act as a buffer, it maintains the pH of the solution at the optimum point for effective washing action. (See Chapter 11 for a discussion of pH and buffers.)

Detergents have provoked considerable concern in recent years because they are among the sources of water pollution. This aspect of detergents is discussed in Chapter 18.

Table 15-2. Phosphate contents of some common detergents

Detergent	% P	% $P_3O_8^{-5}$*	% $Na_5P_3O_8$*
Salvo	14.8	35.2	53.5
Tide, Cheer, Gain	12.3	29.3	44.5
Oxydol	11.7	27.8	42.3
Bold	11.4	27.3	41.2
Dreft	10.0	23.8	36.1
Duz	9.7	23.1	35.0
Fab, Ajax, All, Breeze, Punch, Cold Power	8.7	20.7	31.4

*Calculated from the % P given by the manufacturer (1973).

oven cleaners

Grease from cooking that is splattered and baked onto the walls of an oven is often difficult to remove with ordinary detergents or scouring powders. Nevertheless, there are several products on the market that make this chore relatively easy. These materials usually contain strong bases, such as sodium hydroxide, and carry a warning about the possibility of chemical burns if they are spilled on the skin or clothing. Oven cleaners actually convert the grease to soap by the same process described earlier, in the section on soaps. They convert the fat to glycerol and salt of the corresponding fatty acids, which are water-soluble and hence easily removed.

drain cleaners

Drain cleaners also consist of strong bases such as sodium hydroxide. One commercial product contains solid sodium hydroxide flakes mixed with a few chips of aluminum. The aluminum reacts with the base to form hydrogen gas, which causes the solution to bubble. The bubbling action helps to dissolve the crystals.

$$Al + NaOH \longrightarrow H_2 + Na_3AlO_3$$

The bubbling also makes the mixture look "powerful" and appear to be "working hard."

The two most common materials that tend to clog household drains are grease and hair. Strong bases will attack both these materials. The reaction with grease to form soap has been described already. Hair is composed primarily of proteins, complex molecules consisting of several amino acid units (see Chapter 13). These amino acids are connected via amide linkages that are hydrolyzed easily by strong bases.

$$\underset{\text{Hydrolysis of an amide linkage in a protein molecule}}{-\overset{\displaystyle O}{\overset{\|}{C}}-NH- \quad \xrightarrow{\text{NaOH}} \quad -\overset{\displaystyle O}{\overset{\|}{C}}-O^{-1}Na^{+1} + H_2N-}$$

Once these linkage are broken, the proteins and thus the hair lose structural strength and can be washed away, thereby clearing the drain.

household bleaches

Common liquid household bleach is a 5% solution of sodium hypo-chlorite (NaOCl) in water. A 16% solution is sold for home swimming pools and commonly is called *swimming pool chlorine*. This solution is too strong for normal household purposes.

Sodium hypochlorite may be made by passing an electrical current through a solution of sodium chloride in water.

$$2\ NaCl + H_2O \xrightarrow{e^-} NaOCl + H_2$$

This is the same reaction that is used to make chlorine gas, except there is no attempt to separate the anode and cathode compartments in this process. (See Chapter 10 for a discussion of chlorine generation.) An alternate procedure frequently used in the laboratory for preparing sodium hypochlorite involves the addition of chlorine gas to a solution of sodium hydroxide in water.

$$2\ NaOH + Cl_2 \rightarrow NaOCl + NaCl + H_2O$$

Sodium hypochlorite solutions are unstable and slowly decompose on standing. For this reason, they should be purchased fresh and used as soon as possible. Bleaches in powdered form are preferred sometimes, since they can be stored longer without losing their capacity for chemical activity. Ordinary bleaching powder is a calcium hypochlorite salt (CaCl(ClO)), commonly called *chloride of lime*. A stronger material (Ca(ClO)$_2$) is sold for swimming pool use.

Bleaches work because of their ability to react with carbon-

carbon double bonds. Most colored organic substances owe their color to the presence of several double bonds in the molecule. Both chlorine and hypochlorite can react with double bonds and effectively destroy the color. The chlorinated products are almost always colorless.

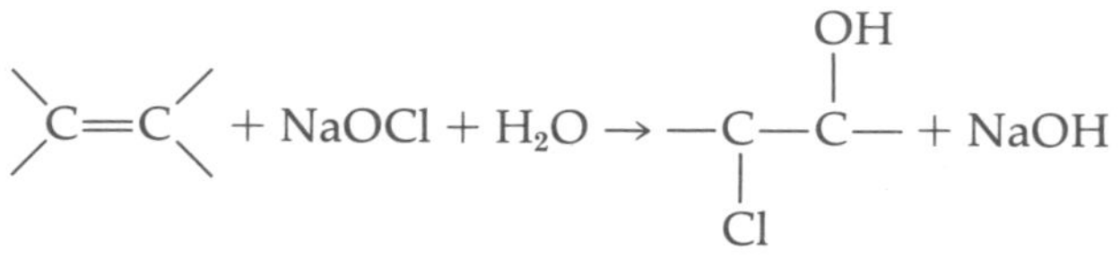

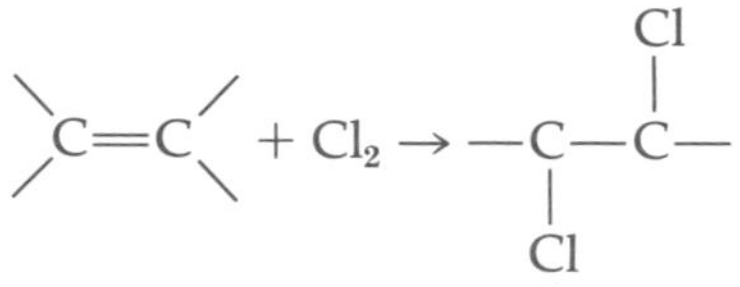

$$\text{C=C} + NaOCl + H_2O \rightarrow -\overset{\displaystyle |}{\underset{\displaystyle |}{C}}-\overset{\displaystyle OH}{\underset{\displaystyle |}{C}}- + NaOH$$

$$\text{C=C} + Cl_2 \rightarrow -\overset{\displaystyle Cl}{\underset{\displaystyle Cl}{C}}-\overset{\displaystyle Cl}{\underset{\displaystyle |}{C}}-$$

hard water and water softeners

When water comes in contact with air, as in most natural situations, some of the gases in the air dissolve in the water. One of these gases, oxygen, makes marine life possible. Another important gas, carbon dioxide, reacts with water to form carbonic acid.

$$H_2O + CO_2 \rightarrow . \quad H_2CO_3$$

Carbonic acid

Carbonic acid partially dissociates to form hydrogencarbonate and carbonate ions in solution and is thus a weak acid.

$$H_2CO_3 \rightarrow H^{+1} + \qquad HCO_3{}^{-1}$$

Hydrogencarbonate ion

$$HCO_3{}^{-1} \rightarrow H^{+1} + \quad CO_3{}^{-2}$$

Carbonate ion

As a consequence of this action, most water is slightly acidic. Typically, water can have a pH as low as 4.5.

When water with dissolved carbon dioxide passes over limestone, marble, or other rocks containing calcium or magnesium carbonates, these metal ions are dissolved in the water.

$$CaCO_3 \qquad + H^{+1} \rightarrow Ca^{+2} + HCO_3{}^{-1}$$

Limestone or marble

Water containing dissolved calcium or magnesium ions is called **hard water**. Most groundwater contains at least some dissolved metal ions. Some water contains so much dissolved material that it develops a distinctive, undesirable taste. This is particularly true in arid parts of the country such as the Southwest. Table 15-3 shows some typical values for the hardness of water in selected cities in the United States.

Table 15-3. Hardness of water in selected cities*

City	Hardness Expressed as $CaCO_3$ (in ppm**)
Birmingham, Alabama	7–95
Phoenix, Arizona	176–418
Tucson, Arizona	119–144
Los Angeles, California	81–323
San Francisco, California	3–111
Denver, Colorado	5–44
Washington, District of Columbia	129–137
Chicago, Illinois	117–133
Boston, Massachusetts	10–13
Detroit, Michigan	96–100
St. Louis, Missouri	77–129
Albuquerque, New Mexico	20–161
New York, New York	18–307
Cleveland, Ohio	44
Dallas, Texas	80–166
Houston, Texas	14–92
Salt Lake City, Utah	136–274
Seattle, Washington	14–30

*C. N. Durfor and E. Becker, "Public Water Supplies of the 100 Largest Cities in the United States, 1962." U.S. Geological Survey Water-Supply Paper 1812.
**ppm = parts per million by weight

In addition to taste problems, hard water also interacts with soaps and detergents to form an insoluble scum.

$$2\ C_{17}H_{35}COONa + Ca^{+2} \rightarrow (C_{17}H_{35}COO)_2Ca + 2\ Na^+$$

Sodium stearate   Calcium stearate
(a typical soap)   (a typical soap scum)

A bathtub ring usually consists of this type of scum. Once the soap has precipitated, it loses its cleaning ability. Thus hard water requires more soap and is not so good as soft water for washing. Detergents have a reduced tendency to form an insoluble scum and are thus preferred by most housewives. In fact, most of the material commonly called *soap* today is actually detergent.

Several techniques have been developed for softening water, or removing the undesirable metal ions from the water. When dissolved carbon dioxide is the only source of acidity, the water is called **temporary hard water,** and the ions can be precipitated as insoluble carbonates by boiling.

$$Ca(HCO_3)_2 \rightarrow CaCO_3 + CO_2 + H_2O$$

$$Mg(HCO_3)_2 \rightarrow MgCO_3 + CO_2 + H_2O$$

This reaction is responsible for the boiler scale found on the inside of a teakettle or a water heater that has been used for several years. Boiler scale conducts heat poorly, causing the efficiency of an old water heater to be lower than that of a new one.

In some cases, water contains negatively charged ions other than hydrogencarbonate ions. Chloride and sulfate ions are typical examples and may enter water from natural, mining, or industrial sources. Water containing these ions is called **permanent hard water** because it cannot be softened by heating. Permanent hard water can be softened by adding sodium carbonate, which is commonly called soda ash or washing soda. Addition of sodium carbonate causes hard water cations to be precipitated as their carbonate salts.

Boiler scale deposited in a water heater. This scale consists mainly of calcium and magnesium carbonates that precipitate from hard water. Heating efficiency is reduced by such deposits. *(Courtesy Culligan Water Institute)*

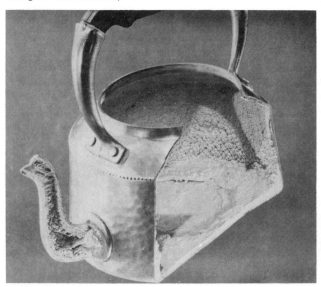

$$Mg^{+2} + CO_3^{-2} \rightarrow MgCO_3$$

$$Ca^{+2} + CO_3^{-2} \rightarrow CaCO_3$$

The original negatively charged ion and the sodium ion remain in solution, but do not precipitate soap, and are thus considered harmless.

The most satisfactory method of reducing hardness in water involves use of a commercial water softener. Commercial water softeners function by a process known as ion exchange. In this process, water passes through a bed of cation-exchange resin or zeolite. A cation-exchange resin is an insoluble polymer of polystyrene or other material that has sulfonic acid groups attached to it. These are usually molded into small beads. A zeolite is a similar inorganic material. The resin is activated by reaction with sodium carbonate or sodium hydroxide, which converts it to the corresponding sodium salt. Figure 15-2 shows an activated cation-exchange resin particle.

Calcium and magnesium ions are held more tightly by the resin than are sodium ions. Thus when hard water is passed through a bed of this resin, the hard-water ions are held by the resin, and the sodium ions are released into the solution. Figure 15-3 illustrates this procedure in a commercial water softener. Over a period of time, the zeolite resin becomes saturated with hard-water ions. Fortunately, the resin can be regenerated by passing a concentrated brine solution (sodium chloride) through it. The effluent wash solution containing the hard-water ions is then discarded.

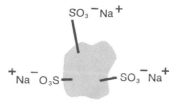

Figure 15-2. A typical cation-exchange resin particle activated with sodium ions

$$\left\{\begin{matrix}-SO_3^{-1}\\-SO_3^{-1}\end{matrix}\right. Ca^{+2} + 2Na^{+1} \rightarrow \left\{\begin{matrix}-SO_3^{-1}Na^{+1}\\-SO_3^{-1}Na^{+1}\end{matrix}\right. + Ca^{+2}$$

Special resins have been developed that will exchange hydrogen ions for cations and hydroxyl ions for negatively charged ions. When these resins are used together, it is possible to reduce the ionic impurity content of the water to very low levels.

$$^{\oplus}\left\{OH^{-1} + Cl^{-1} \rightarrow {}^{\oplus}\right\}Cl^{-1} + OH^{-1}$$

$$H^{+1} + Na^{+1} \rightarrow {}^{\ominus}\left\{Na^{+1} + H^{+1}\right.$$

$$H^{+1} + OH^{-1} \rightarrow H_2O$$

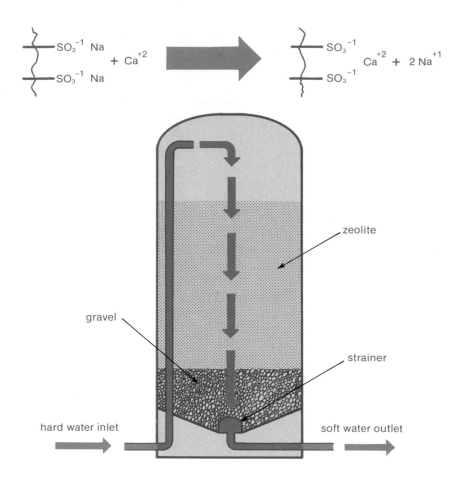

Figure 15-3. A commercial
water softener

Water treated in this manner is called **demineralized water**, because
the mineral content has been removed. A small inexpensive unit has
been marketed, employing ion-exchange resins, that removes undesirable ions from water to be used in steam irons.

study questions

1 Why do people cook food?
2 What is the difference between a fat and an oil?
3 What chemical compound causes the eye irritation associated with frying food?

4 In addition to providing flavor, what is the function of salt in bread?

5 What is the effect of yeast on sugar?

6 How might you make baking powder if the commercial product were unavailable?

7 What is the difference between a soap and a detergent?

8 How might you make soap at home?

9 What are aluminum soaps, and for what are they used?

10 How does soap clean?

11 What is the function of phosphate in a detergent?

12 What causes hard water?

13 What causes boiler scale in a teapot?

14 What is soap scum?

15 How does a commercial water softener work?

special problems

1 Discuss the pollution problems associated with some of the household products discussed in this chapter (for example, detergents, drain cleaners, oven cleaners, and bleaches).

2 Discuss the energy requirements of some of the household products discussed in this chapter.

plastics and synthetic fibers

The plastics and synthetic fibers industry has grown from almost nothing at the turn of the century to a multibillion dollar industry, employing thousands of people in the United States and thousands more in the rest of the world. During this time, plastics and synthetic fibers have essentially revolutionized our society. Some individual plastics like nylon, polyvinylchloride (PVC), and polyethylene are produced in excess of a billion pounds per year. Many manufactured products that formerly were made from wood or metal now are made from plastics. These include building materials, automotive parts, and machinery. In many cases, the plastic item is of higher quality and lower cost than the equivalent product made from other materials. In some cases, such as electrical plugs and receptacles, there is no really acceptable alternative material. Products such as today's clothes, furniture, toys, cars, and homes would be grossly different in style, durability, and cost if they were not made at least partially from plastics.

A few plasticlike materials occur naturally, including starch, cellulose, natural rubber, and proteins. Most plastics used today, however, are synthetic materials, and nearly all have been developed within the past 30 or 40 years. This chapter gives a brief look at the chemistry of these new materials.

polymer
thermoplastic polymer
thermosetting polymer
elastomer
monomer
polymerization
addition polymerization
condensation polymerization
vulcanization

polymers

Plastics and synthetic fibers consist of large organic molecules called polymers. The molecular weights of polymer molecules frequently range between several thousand and several hundred thousand atomic mass units. Polymers are divided into three general types, depending on their physical properties—thermoplastic polymers, thermosetting polymers, and elastomers.

As an example of the extent to which new applications for plastics are being found, Merck Sharp and Dohme Orthopedics Company is marketing a new plastic material for setting broken bones. The new plastic cast has an outer layer consisting of a slightly sticky glass-fiber tape. Exposure to ultraviolet light causes a photosensitive resin to set into a hard plastic. The new material is reported to be stronger, faster curing, lighter weight, and easier to remove than the conventional plaster of paris cast.

Thermoplastic polymers will melt on being heated and can be molded easily into new shapes. Most of the common plastics such as polyvinylchloride (PVC), polyethylene, and nylon are of this type. Synthetic fibers also are made from thermoplastic polymers.

When heated, thermosetting polymers slowly char rather than melt. They must be made in their final shape or machined into shape because they cannot be molded. This property makes thermosetting plastics useful for electrical insulation and handles for pots and pans. Bakelite is the best-known example of this type of polymer; it is used widely in the electrical industry as an insulating material. The Glyptals are also thermosetting polymers.

Elastomers are rubbery or elastic materials. Natural rubber and the various synthetic rubbers such as neoprene are elastomers. They are used in such commercial products as automobile tires, toy balloons, and shock-absorbing devices.

In addition to the basic polymer, most plastics also contain fillers, plasticizers, and dyes or pigments. Fillers such as diatomaceous earth, wood pulp, and cotton lower the cost and modify the properties of the plastic. Plasticizers make the plastic harder, more flexible, or more easily molded. Polyvinylchloride (PVC), for example, is a rigid, solid substance. When the appropriate plasticizers are added, it becomes soft and flexible, making it useful for inflatable toys and flexible tubing. Similarly, pure polypropylene has little commercial value, but with plasticizers added, it is used extensively for plastic containers that require hinges. Polypropylene hinges normally can be operated thousands of times before they break.

monomers

Polymers are made from small molecules called monomers. Most monomers contain between two and eight carbon atoms. There are two major types of monomers. The first type contains a carbon-

carbon double bond and one other side group. This type of monomer usually conforms to the formula $CH_2\!=\!CH\!-\!X$, where X is a side group. Examples of this type of monomer are: ethylene, which is used to make polyethylene; propylene, which is used to make polypropylene; styrene, which is used to make polystyrene; and vinyl chloride, which is used to make polyvinylchloride.

The second type of monomer contains a carbon skeleton with a functional group at each end. Some examples of this type are: adipic acid and hexamethylenediamine, which are used to make nylon 66; terephthalic acid, which is used in polyester fibers; and epichlorohydrin, which is used in epoxy resins.

$$CH_2\!=\!CH$$
$$|$$
$$H$$

Ethylene

$$CH_2\!=\!CH$$
$$|$$
$$CH_3$$

Propylene

$$HOOCCH_2CH_2CH_2CH_2COOH$$
Adipic acid

$$CH_2\!-\!CHCH_2Cl$$
$$\diagdown O \diagup$$

$$H_2NCH_2CH_2CH_2CH_2CH_2CH_2NH_2$$
Hexamethylenediamine

Terephthalic acid

Epichlorohydrin

$$CH_2\!=\!CH$$

Styrene

As might be expected, the monomer industry, like the plastics and synthetic fibers industry, is a multibillion dollar industry. Owing to extensive research and development, commercial production of monomers has reached a high level of sophistication. Consequently, prices of polymers generally have declined, even though prices of most other commercial products generally have increased. As a rule, monomers are made from petroleum hydrocarbons, and thus their future price depends on the price and availability of petroleum. Monomers are now usually made in huge industrial complexes that are each capable of producing 300 million pounds or more of material per year.

$$CH_2\!=\!CH$$
$$|$$
$$Cl$$

Vinyl chloride

addition polymers

The process of converting a monomer into a polymer is known as **polymerization**. Polymers are made by combining monomers into long chains of various lengths. The first type of monomer (those which conform to the general formula $CH_2\!=\!CH\!-\!X$) undergo **addition polymerization**. Addition polymerization takes place when monomer groups are successively added together to form a polymer. Some examples of addition polymers and their uses are given in Table 16-1.

In 1974 the B. F. Goodrich Company disclosed that three employees at its vinyl chloride plant in Louisville, Kentucky, had died of angrosarcoma of the liver. Shortly after this disclosure, several other companies reported similar observations. About this same time tests on laboratory animals exposed to vinyl chloride vapors also revealed an unusually high incidence of this rare form of cancer. The Environmental Protection Agency and the companies manufacturing vinyl chloride and vinyl chloride products acted quickly to minimize further exposure of workers and the general public to this chemical, which was formerly believed safe enough to use as a propellant in aerosol spray cans.

$$\cdots + \overset{}{\underset{X}{CH_2{=}CH}} + \overset{}{\underset{X}{CH_2{=}CH}} + \overset{}{\underset{X}{CH_2{=}CH}} + \cdots \rightarrow$$

$$\cdots \overset{}{\underset{X}{CH_2{-}CH}} \overset{}{\underset{X}{CH_2{-}CH}} \overset{}{\underset{X}{CH_2{-}CH}} \cdots$$

Polyethylene is a particularly interesting product because there are two major processes for its manufacture. The older process, known as the *free-radical process*, produces polymer molecules with

Table 16-1. Selected addition polymers and their uses

Monomer	Polymer	Use
$CH_2{=}CH_2$ Ethylene	Polyethylene	Plastic bags, trash containers, toys
$CH_2{=}CHCH_3$ Propylene	Polypropylene	Fibers, containers requiring hinges
$CH_2{=}CH{-}Cl$ Vinyl chloride	Polyvinylchloride	Film, inflatable toys, plastic pipe
$CH_2{=}CH{-}\bigcirc$ Styrene	Polystyrene	Molded items, foamed packaging
$CH_2{=}CCH_3COOCH_3$ Methyl methacrylate	Polymethyl methacrylate, Lucite, Plexiglas	Coatings, clear plastic sheets
$CH_2{=}CClCH{=}CH_2$ Chloroprene	Neoprene	Rubber goods
$CF_2{=}CF_2$ Tetrafluoroethylene	Teflon	Coatings on cooking utensils, gaskets, insulation

Over the years, research has led to more efficient processes for preparing monomers, using less and less expensive raw materials. For example, acrylonitrile is now made from propylene, air, and ammonia in a process developed by the Standard Oil Company (Ohio).

$$CH_3CH = CH_2 + O_2 + NH_3 \rightarrow CH_2 = CH - CN$$

An older process for making acrylonitrile was to treat acetylene with hydrogen cyanide.

$$HC \equiv CN + HCN \rightarrow CH_2 = CH - CN$$

The new process for manufacturing acrylonitrile is not only less expensive, but also less hazardous for plant workers because hydrogen cyanide is not used as a reactant (some hydrogen cyanide is produced as a by-product).

---

extensive chain-branching and random-chain length. The polymer molecules from this process fit together poorly, giving the resulting plastic a low density, low melting point, and low mechanical strength. Polyethylene prepared in this manner is called *low-density polyethylene*.

The more recent process for making polyethylene is called the *Ziegler-Natta process*, after the two discoverers of the basic catalyst required. The product from the Ziegler-Natta process tends to contain linear polymer molecules of more uniform chain length than those obtained by the free-radical process. The resulting polymer is more dense, mechanically stronger, and has a higher melting point than low-density polyethylene. Polyethylene manufactured in this manner is called *high-density polyethylene*.

condensation polymers

The second type of monomer undergoes condensation polymerization. In this process the functional groups on one monomer molecule react with the functional groups on another, to make a single polymer molecule. Polyester fibers are examples of this type of polymer. A typical polyester is made by condensing dimethyl terephthalate with ethylene glycol.

| Dimethyl terephthalate | Ethylene glycol | a polyester | + n CH₃OH |

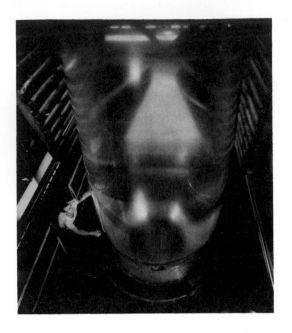

Left above: Two-story high bubble is formed during extrusion of polyethylene film. This film will be used to make thousands of plastic bags for packaging. It also can be used industrially as a vapor and moisture barrier and agriculturally for mulching, silage covers, greenhouse glazings, pond liners, and animal shelters. *(Courtesy Union Carbide Corporation)*

Right above: Film of neoprene (synthetic rubber) in final stages of manufacture. *(Courtesy E. I. du Pont de Nemours & Co., Wilmington, Delaware)*

Polyesters are so called because the functional group connecting the individual monomer units is an ester group.

The nylons are another group of condensation polymers. They are sometimes called *polyamides* because they are held together by amide groups. There are two types of nylon in common use, nylon 66 and nylon 6. Nylon 66 is made by condensing adipic acid with hexamethylenediamine.

$$HOOC(CH_2)_4COOH$$
$$+$$
$$H_2N(CH_2)_6NH_2$$

$$\rightarrow \left[\underset{\parallel}{\overset{O}{C}}(CH_2)_4\underset{\parallel}{\overset{O}{C}}-NH(CH_2)_6NH \right]_n$$

Nylon 66

The first 6 in the name *nylon 66* represents the number of carbon atoms in one monomer unit, and the second 6 represents the number of carbon atoms in the second monomer unit.

Nylon 6 is made by heating caprolactam to induce polymerization.

Caprolactam $\rightarrow \left[NH-CH_2CH_2CH_2CH_2CH_2C \right]_n$ Nylon 6

This product is called *nylon 6* because there is only one monomer unit, and it contains six carbon atoms. Nylons are used extensively in textile fibers and for molded parts requiring high strength. Table 16-2 lists some condensation polymers and their uses.

Nylon was the first entirely synthetic fiber. Its synthesis as a textile fiber was reported in 1938, about eight years after its initial discovery. Nylon thread is formed by melting crude nylon polymer and forcing the thick liquid through tiny holes in discs (spinnerettes). As the nylon emerges from these discs, it solidifies into fibers that are woven into strong, elastic thread and yarn. Molten nylon polymer also can be poured into molds for the manufacture of many items.

Molten nylon being extruded through small holes in a spinneret. The fibers thus formed are woven into thread for subsequent use. *(Courtesy Celanese Fibers)*

naturally occurring polymers

Naturally occurring polymers obtained from plants and animals have been known for hundreds of years. Resins, gum rosin, and rubber are a few such substances obtained from plants. Shellac is a polymeric substance that is secreted by an insect. Cellulose and proteins also are polymers found in plants and animals.

Natural rubber is obtained from *Hevea brasiliensis*, a tree indigenous to South America, and has been used by the natives of the area for several centuries. It was not introduced into Europe, however, until after Cortez observed the Aztec Indians playing with rubber balls around 1520. Rubber found only limited use until the middle 1800s, because of its tendency to become sticky in warm weather and brittle in cold weather. In 1839, Charles Goodyear found that rubber could be treated with sulfur to form a more useful product. The discovery of this process, known as vulcanization, was the beginning of the rubber industry and the manufacture of many rubber products.

By World War II, the need for rubber had become great. At the same time, the major supplies of natural rubber to the United States

---

Goodyear's discovery of the vulcanization process occurred as the result of an accidental spilling of a sulfur-rubber mixture on his hot stove, which charred the material into a leathery substance. William Brockedon, an English inventor who saw samples of Goodyear's sulfur-treated rubber, named the process "vulcanization" after Vulcan, the mythological god of fire. Today, many rubber goods are molded under heat and pressure and are thus shaped and vulcanized at the same time.

Table 16-2. Some condensation polymers and their uses

Monomer	Polymer	Use
CH$_3$OOC—⟨benzene ring⟩—COOCH$_3$ Dimethylterephthalate and CH$_2$CH$_2$ | | OH OH Ethylene glycol	Polyester Dacron Mylar	Clothing, film
⟨Caprolactam ring structure with O and N—H⟩ Caprolactam	Nylon 6 Perlon	Fibers, molded items
HOOC(CH$_2$)$_4$COOH Adipic Acid and HN$_2$(CH$_2$)$_6$NH$_2$ Hexamethylenediamine	Nylon 66 Zytel	Fibers, molded items
⟨Glucose ring structure with CH$_2$OH, OH, HO, OH, OH⟩ Glucose	Starch	Food
	Cellulose	Wood, rayon, cellophane
Assorted amino acids*	Protein	Muscle tissue of living organisms, enzymes

*See Figure 13-5 for a list of naturally occurring amino acids.

were cut off by the Japanese. The United States government then mounted a tremendous research and development effort, which resulted in the discovery of Buna S rubber. Buna S was made by polymerizing styrene and butadiene.

Both starting materials could be obtained from petroleum, which was readily available within the United States. The structure of Buna S significantly differs from that of natural rubber, which may be thought of as a polymer of isoprene.

$$\text{Styrene} + CH_2{=}CH{-}CH{=}CH_2 \rightarrow \text{Buna S}$$

Styrene   Butadiene   Buna S

Isoprene   Natural rubber

Despite the difference in structures, the properties of the new synthetic materials were adequate for commercial use.

Since World War II, many new types of synthetic rubbers have been developed and commercialized. These new materials have properties superior to the older products, and in some cases very nearly approximate natural rubber in both structure and physical properties.

Cellulose and starch are two other closely related, naturally occurring polymers. Living organisms manufacture them by condensing glucose molecules into polymeric chains. Two synthetic materials, rayon and cellophane, are made from cellulose. Rayon is made by treating wood pulp with sodium hydroxide followed by carbon disulfide. The resulting viscous mass then is passed through spinnerettes into a bath of sulfuric acid, which regenerates the cellulose. The resulting threads are bleached, dyed, and woven into cloth. In

Left: Wood engraving showing early apparatus used in making vulcanized rubber tires for bicycles and coaches. *(Courtesy Uniroyal, Inc., New York)*

Right: Rubber-moulding operation. Rear tractor tires in curing presses where vulcanization occurs. *(Courtesy Goodyear)*

the early 1950s, rayon was probably the most popular synthetic fabric material. It has since been largely displaced by newer synthetic fibers. Cellophane has the same chemical composition as rayon, except that it is made into sheets rather than fibers.

disposal problems

As the use of plastics and synthetic fibers has increased, the problem of disposal also has increased. Most plastics are essentially inert to the normal degradation processes of nature—that is, they are *not* biodegradable. It is thought by some that many plastics will retain their form in sanitary landfills for a thousand years or more without being destroyed. This longevity can create a tremendous solid-waste disposal problem over many years. The magnitude of this problem has led many people to propose that use of plastics be greatly curtailed. Use of plastics in packaging has been criticized especially, because plastic packages normally are used only once and then discarded.

One solution would be to reuse the packaging. According to an article in *Fortune* (Tom Alexander, "The Packaging Problem Is a Can of Worms," June 1972), people *say* that they favor recycling, but in actual practice they are reluctant to recycle, even when refusal costs them more money. As evidence for this claim, the increased use of nonreturnable soft-drink containers during the past decade was

Reusable packaging is one solution to garbage-disposal problems. *(Photograph by A. Marshall Licht)*

cited. Furthermore, even when returnable containers are purchased, there is an increasing tendency not to return them. A few years ago, the common soft-drink bottle had an average lifetime of about 40 cycles. More recently the average has dropped to 15 cycles and is as low as 4 in some urban areas. A returnable bottle must be reused at least 6 times to make up the difference in cost between the returnable and nonreturnable bottle. It is no wonder, then, that the soft-drink bottlers favor nonreturnable bottles.

To meet the demand for lower cost, throwaway bottles, some companies are developing plastic soft-drink bottles. Proponents of plastic bottles point out that they are safer than glass ones because they are unbreakable.

Arguments favoring plastic-film packaging of produce also were put forward in the *Fortune* article. It has been estimated that losses for such items as unpackaged lettuce and grapes run as high as 20%. Losses are reduced to about 2% when the produce is wrapped in plastic film. Since plastic film is inexpensive, there is a significant economic advantage to wrapping the produce in plastic film, helping to reduce the cost of food. Furthermore, the reduction in food spoilage reduces the garbage-disposal problem. It should be pointed out, however, that garbage is biodegradable, but plastic film generally is not biodegradable. Consequently, the pollution problem from plastic film is longer lasting. The question of whether to use no packaging and suffer high produce losses or to use nondegradable packaging and suffer lower losses is one that still is debated. Meanwhile, re-

Plastic-film packaging of produce reduces overall food losses, but itself contributes to garbage-disposal problems. *(Photograph by A. Marshall Licht)*

search is underway to develop new types of packaging that would be both effective and degradable. Such materials are not yet commercially available, however.

Environmentalists are pushing for laws that would make many current packaging methods unlawful or too expensive to use. Generally, manufacturers are resisting these efforts. This is not an area where it is obvious which side is right and which is wrong. It is an area, however, where decisions will profoundly affect the future of society. Pollution problems are discussed further in Chapters 17 and 18.

study questions

1 From what are plastics made?
2 Give three examples of products made with a thermosetting plastic.
3 Give three examples of products made with a thermoplastic plastic.
4 What is the function of additives in plastics?
5 When petroleum becomes scarce, plastic goods can be expected to become scarce also. Why?
6 By what two principal chemical reactions may polymers be synthesized?
7 Is it true that plastic products are not so good as equivalent products made from other materials?
8 By what process can natural rubber be made more durable for use in hot and cold weather?
9 Name some natural polymers.
10 Do you favor the use of synthetic (plastic) food packaging? Why?
11 It has been suggested that some polymers could remain unchanged for more than a thousand years in a sanitary landfill. What harm would this cause?
12 How does low-density polyethylene differ from high-density polyethylene?

special problems

1 Suggest a structure for a polymer made from the following.

a. [benzene ring with two $-CO_2H$ groups] and $HO-CH_2CH_2CH_2CH_2-OH$

b. [benzene ring with CO_2H at top and CO_2H at bottom] and $H_2N-CH_2CH_2CH_2CH_2-NH_2$

2 To what known types of synthetic polymers are the products of the reactions in Problem 1 related?

3 Suggest a structure for a polymer made from the following.
   a. Styrene   b. Vinyl chloride

4 When a polymer is formed from chloroprene, three possible geometric isomers may result: an all-*cis*, an all-*trans*, or a random *cis-trans* geometric isomer. Draw structures to illustrate these three possible products. (Refer to Chapter 12 for a review of geometric isomers.)

chapter **17**

chemistry and the atmosphere

The importance of the chemistry of the environment has been only recently recognized by the general public. Just a few years ago, it was believed that only Los Angeles had a serious air-pollution problem. Today every major city in the world is openly faced with this problem. Water pollution has also been largely ignored until recently, even though it has been recognized as a serious problem in a few parts of the world for over 200 years.

For the purpose of the following discussion, pollution is defined as the presence of one or more chemical substances at an excessively high concentration somewhere in the environment. It must be recognized that many factors are involved in determining what constitutes an excessively high concentration. Pollutants may be toxic or objectionable materials, such as mercury compounds or sulfur dioxide, or they may be materials such as phosphates from detergents and nitrates from fertilizers, which, in moderate amounts, are necessary for plant growth. In excess amounts, however, these latter materials accelerate the growth of algae in rivers and lakes so that fish die owing to lack of oxygen.

Thermal pollution was discussed in the chapter on nuclear chemistry (Chapter 5) and is not included here.

pollution
aerosol
fly ash
coal gasification
gasoline
octane rating

air pollution

When one thinks of air pollution, one normally thinks of man-made pollutants. Actually, air pollution also results from natural phenomena. Winds blow dust into the air. Natural phenomena like volcanoes and forest fires put ashes, soot, carbon dioxide, sulfur dioxide, and other materials into the air. It has been shown that the natural haze in such areas as the Blue Ridge Mountains of Virginia is caused by submicroscopic particles in the air. These particles are known as

Los Angeles Civic Center area. *Left*: A clear day. *Right*: A smoggy day. *(Courtesy County of Los Angeles, Air Pollution Control District)*

aerosols. In this case, the aerosols are believed to be derived from the naturally occurring compounds found in trees. Dust and smoke are also sources of aerosols.

It has been estimated that 1×10^{12} tons of material enter the atmosphere each year. Of this, only 5×10^8 tons, or about 0.05%, is due to the activities of modern man. That portion, however, is frequently the most objectionable, leading to reduced visibility, eye and nose irritation, health problems, and property damage. Nearly every activity of modern man leads either directly or indirectly to some form of air pollution.

Most chemicals that end up in the atmosphere are fairly simple. Unfortunately, complex chemical reactions often take place in the atmosphere that convert many simple pollutants into more objectionable materials. The energy for these reactions comes primarily from sunlight. Thus, heavily industrialized cities in the warmer or sunnier areas of the world tend to have the most severe air-pollution problems. Los Angeles was one of the first major cities of the United States to experience major air-pollution problems, because of its unique geographic condition and the frequency of hot, sunny days there.

The chemistry of air pollution is complex and only partially understood, even after several years of intensive study. Nitrogen oxide, sulfur dioxide, carbon monoxide, and organic compounds (primarily hydrocarbons) are among the important pollutants added to the air by man. These materials undergo light-induced reactions to form such substances as ozone, aldehydes, and peroxyacyl nitrates. Some of these will be discussed individually in the next few sections.

The solution to the smog problem must involve a reduction in all pollutants entering the atmosphere. It has been shown, for example, that a reduction in just the hydrocarbons in one location can lead to an increase in the smog levels of nearby areas that contain natural sources of hydrocarbons, such as pine trees. This phenomenon occurs because ozone, generated from nitrogen dioxide and sunlight, builds up to higher levels than before in the first area and drifts to nearby areas before it can react with available hydrocarbons.

sunlight and air pollution

Nitric oxide is a by-product in the automobile engine and in certain industrial processes. (These sources will be discussed later.) In the presence of air and sunlight, nitric oxide is converted to nitrogen dioxide.

$$2 \text{ NO} + \text{O}_2 \xrightarrow{\text{sunlight}} 2 \text{ NO}_2$$

Nitric   Nitrogen
oxide   dioxide

Nitrogen dioxide is believed to decompose in the presence of sunlight to produce an oxygen atom and to regenerate nitric oxide.

$$\text{NO}_2 \xrightarrow{\text{sunlight}} \text{NO} + \text{O}$$

Oxygen
atom

In this manner, a photochemical cycle is established that produces free oxygen atoms.

$$\text{O}_2 + 2 \text{ NO} \rightarrow 2 \text{ NO}_2$$
$$\uparrow \qquad \downarrow$$
$$2 \text{ NO} + 2 \text{ O}$$

Free oxygen atoms are much more reactive than oxygen molecules. It is believed that oxygen atoms combine with oxygen molecules in the air to form ozone.

$$\text{O}_2 + \text{O} \rightarrow \text{O}_3$$

Ozone

Ozone is a highly reactive form of oxygen that is responsible for the rapid deterioration of rubber articles, such as tires, in industrial cities. Ozone concentration is relatively easy to measure, and its quantity in the air is used as an indicator of the amount of photochemical smog present. Ozone also is produced naturally during lightning storms by the interaction of electrical discharges with oxygen in the air.

Oxygen atoms react with hydrocarbons in the air to form highly reactive species called *free radicals*. These free radicals then react with other substances in the air, such as oxygen and nitrogen oxide, to produce many other irritants of smog.

sulfur dioxide

Sulfur dioxide is an air pollutant that deserves special attention. It enters the atmosphere from both natural and man-made sources. The major man-made source has been the combustion of sulfur-bearing fuels, principally unrefined petroleum and coal. Certain metal ores are also sources of sulfur dioxide. Techniques for separating the sulfur from crude petroleum are now reasonably well developed; as a result, refined petroleum is no longer the major source of sulfur dioxide pollution.

It is more difficult to remove sulfur from coal before it is burned. Consequently, coal combustion is presently the major source of sulfur dioxide pollution. The growing emphasis on coal as a source of energy for generating electrical power and "synthetic" natural gas is increasing the sulfur dioxide pollution problem.

Sulfur dioxide is a particularly undesirable pollutant because it causes eye irritation and plant damage. There is also reason to believe that it causes, or at least aggravates, certain other health problems. The damaging effects of sulfur dioxide are probably due to its ability to form sulfurous acid with water.

$$SO_2 + H_2O \rightarrow \quad H_2SO_3$$
Sulfurous acid

Sulfur dioxide also can be oxidized in the air to sulfur trioxide, which forms sulfuric acid when it comes in contact with moisture.

$$2\,SO_2 + O_2 \rightarrow \quad 2\,SO_3$$
Sulfur trioxide

$$SO_3 + H_2O \rightarrow \quad H_2SO_4$$
Sulfuric acid

The process by which this oxidation reaction takes place is not defi-

European linden trees at a busy intersection in the Bronx, New York, show harmful effects probably due to pollutants in automobile and bus exhausts. *(Courtesy Environmental Protection Agency)*

nitely known. One suggestion is that oxidation occurs on the surface of dust particles in the air.

Sulfuric acid and sulfurous acid both undergo the typical reactions of acids. (Acids and bases were discussed in Chapter 11.) They are responsible for much of the corrosion of metal parts exposed to the atmosphere. They also are responsible for the accelerated decay of stone statues and buildings observed in recent years. Many of the famous old statues in Europe have been badly damaged in recent years by high concentrations of sulfuric acid in the atmosphere. Many of these statues were constructed from marble or limestone, which consist primarily of calcium carbonate and are especially vulnerable to attack by acids.

$$CaCO_3 + H_2SO_4 \rightarrow CaSO_4 + H_2O + CO_2$$

Marble or   Calcium
limestone   sulfate

According to electrical power companies, at present no processes for removing sulfur dioxide from the exhaust stacks of large power plants have been proved commercially practical. Several processes are under development, however. One of these processes involves additions of limestone to the furnace along with the coal. In the furnace the limestone is converted to calcium oxide by the heat.

Left: Statue of Alexander Hamilton in Washington, D.C., shows damage due to air pollution. *(Courtesy Environmental Protection Agency)*

Right: Corrosive damage of air pollution on truck body. *(Courtesy Environmental Protection Agency)*

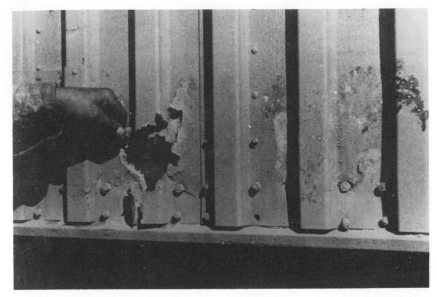

$$CaCO_3 \xrightarrow{\text{heat}} CaO + CO_2$$
$$\text{Calcium}$$
$$\text{oxide}$$

The calcium oxide then reacts with sulfur oxide to form calcium sulfate.

$$CaO + SO_3 \rightarrow CaSO_4$$

This process has the disadvantage that it requires the mining of limestone and the disposing of calcium sulfate.

Other processes under development attempt to produce sulfur or sulfuric acid as salable by-products to partially offset the operating cost of the recovery process. The basic chemistry for these processes is well known. Sulfur dioxide can be reduced to hydrogen sulfide in a stream of partially burned hydrocarbons containing carbon monoxide and hydrogen.

$$3\ H_2 + SO_2 \rightarrow H_2S + 2\ H_2O$$
$$\text{Hydrogen}$$
$$\text{sulfide}$$

The hydrogen sulfide from this operation is then treated with sulfur dioxide to produce elemental sulfur.

$$2\ H_2S + SO_2 \rightarrow 3\ S + 2\ H_2O$$

Petroleum refineries have used this type of technology for years in their desulfurization operations. It has been difficult to adapt it to power plants and copper smelters, however, because the sulfur dioxide concentration is very low in stack gases. Thus the concentration of sulfur dioxide must be increased in order to use this process. It is the concentration step that is the source of the technological problem. The concentrating unit must be sufficiently active that virtually all the sulfur dioxide will be collected. At the same time, it must be capable of handling large volumes of gas. Schemes involving absorption of sulfur dioxide by ammonia, aqueous salt solutions, and even molten salts are presently under investigation.

the electrical power plant problem
in the southwest

Electrical power consumption in the United States has been increasing at a tremendous rate, far exceeding the rate of population growth. Furthermore, as air-pollution levels in cities increase, the demand for "clean" electricity is expected to accelerate. As a consequence of this growth, new power plants are badly needed throughout the country. Unfortunately, although electricity is clean to use, its generation is often the source of major pollution problems.

One of the present problem areas is the southwestern part of the country. Electrical power demands in Los Angeles, Phoenix, Tucson, Las Vegas, and other cities and towns in the Southwest have led to the construction, or proposed construction, of huge coal-burning power complexes in New Mexico, Arizona, Utah, and Nevada. Figure 17-1 shows the locations of some of the existing and planned facilities. Owing to the conflict between energy and environmental concerns, it is uncertain how many of these plants will be constructed.

If all the proposed power plants are constructed, it is expected that the combined capacity will be 36 million kilowatts by 1985. Coal consumption will be nearly 150,000 tons per day. Most of the coal will be strip-mined in the general area, devastating thousands of acres of land. At the same time, these plants will be emitting tremendous quantities of air pollutants unless preventive action is taken.

The principal pollutants currently expected are sulfur dioxide, nitrogen oxides, and fly ash. Sulfur dioxide and nitrogen dioxide have been discussed already. Fly ash is particulate matter and leads to reduced visibility. Fine particles of fly ash (aerosols) may stay airborne for months and travel long distances from the source, but most of the material probably settles out within 100 or 200 miles.

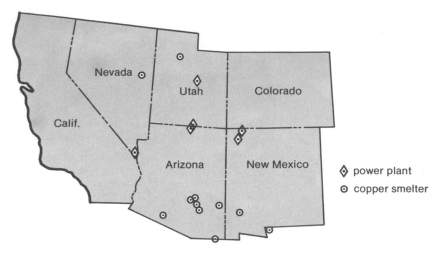

Figure 17-1   Some of the existing and proposed power plants and copper smelters in the Southwest

Heavy smoke-stack emissions can be reduced significantly by using electrostatic precipitators. *(Courtesy Western Precipitation Division, Joy Manufacturing Company)*

The pollution problem created in the Southwest by these plants is complicated further by two other needs of modern man—for copper and for natural gas. Most of the copper produced in the United States is refined in this same area. Copper occurs in several minerals. Two of the most common copper ores are chalcopyrite ($CuFeS_2$) and chalcocite (Cu_2S). The first step in the refining process (called *copper smelting*) involves heating the ore in air to drive off the sulfur as sulfur dioxide.

Above: At night, an electrical power plant casts its reflection of smoke stacks and emissions on the surface of nearby water. *(Courtesy Environmental Protection Agency)*

Left: Open-pit asbestos mine. Strip-mining can devastate sizable portions of land. *(UPI Photo)*

$$4 \ CuFeS_2 + 11 \ O_2 \rightarrow 4 \ CuO + 2 \ Fe_2O_3 + 8 \ SO_2$$

$$Cu_2S + 2 \ O_2 \rightarrow 2 \ CuO + SO_2$$

Normally, the sulfur dioxide escapes into the air. Copper refiners have been reluctant to release information about how much sulfur dioxide is released in each plant because it would be easy for competitors to calculate the production at the plant from this information. (The technique for the calculations is similar to that used on page 140, in Chapter 9.) Nevertheless, it is known that hundreds of tons of sulfur dioxide are produced each day by copper smelters in the area.

Sulfur dioxide from coal gasification is a relatively new problem. Coal gasification is a method of converting coal to synthetic natural gas. (The pertinent chemical processes are discussed in Chapter 19.) The need to replenish dwindling natural gas supplies in the United States in the near future requires construction of coal gasification plants. These plants will consume large amounts of coal and, for economic reasons, must be located near coal deposits. Sulfur dioxide emissions from such plants would add to pollution problems of these areas.

Obviously, the states involved are actively trying to ensure that adequate pollution-control measures are installed in each plant. Unfortunately, adequate pollution-control measures are not always available, as for example with sulfur dioxide or nitrogen oxide. Even with pollution-control devices that can remove up to 99% of the pollutant, as with fly ash, the 1% that is emitted still creates a large pollution problem when several plants are concentrated in a relatively small area.

The problem is complex and will require a great deal of effort to resolve. Two factors seem to be in opposition. More power is needed, yet further pollution of the air will lead to deterioration of the quality of life and will increase health hazards. The problem is not limited to the Southwest, of course. The entire nation is facing an energy crisis. Insufficient power supplies have already caused power shortages and brownouts in the East. These can be expected to continue and probably increase until the problem is resolved. The role of chemistry in the energy crisis is discussed in Chapter 19.

the automobile and its fuel

One of the major sources of air pollution is the automobile. The common fuel for the automobile is gasoline, which is primarily a mixture of hydrocarbons derived from petroleum. In addition, most gasoline contains additives that give it certain desirable properties. The exact composition of gasoline varies, depending on the refinery, the time of year, and the climate. Typically, the major components will be hydrocarbon isomers containing between seven and ten carbon atoms. In cold climates, lower-boiling hydrocarbons such as butanes are added to gasoline for the winter, to increase the volatility of the gasoline and thus make the engine start more easily. During the summer, and in warmer climates, the lower-boiling components are decreased and higher-boiling components are increased, to lower the gasoline's volatility and minimize such problems as vapor lock.

All these changes in composition must be made without significantly affecting the octane rating of the gasoline. The octane rating is a measure of the manner in which the gasoline burns in the engine. A straight-chain hydrocarbon tends to burn with a sharp explosion, which causes the engine to knock or ping and can eventually damage some engine parts. Branched-chain hydrocarbons tend to burn more slowly. Ideally, the gasoline should burn at such a rate that there is steady pressure on the piston during the power part of the cycle. All the fuel should be consumed by the end of the cycle. In actual practice this ideal is not achieved.

In the early days of the automobile industry, it became desirable to have a quantitative measure of the tendency of a fuel to cause an engine to knock. The octane rating resulted from this need. By definition, *n*-heptane is assigned an octane rating of zero, and the compound $(CH_3)_2CHCH_2C(CH_3)_3$ (commonly called *isooctane*) is assigned an octane rating of 100. A high-grade premium gasoline will have an octane rating near 100. The octane ratings of a few typical hydrocarbons are listed in Table 17-1.

Table 17-1. Octane ratings of some selected hydrocarbons

Name	Structure	Octane Rating
Toluene	(benzene ring)—CH_3	120.1
Benzene	(benzene ring)	105.0
Propylene	$CH_2{=}CHCH_3$	102.5
Cyclopentane	$(CH_2)_5$	101.3
1-Butene	$CH_2{=}CHCH_2CH_3$	97.4
Propane	$CH_3CH_2CH_3$	97.1
n-Butane	$CH_3CH_2CH_2CH_3$	93.8
2,4-Dimethylhexane	$(CH_3)_2CHCH_2CHCH_3CH_2CH_3$	65.2
n-Hexane	$CH_3(CH_2)_4CH_3$	24.8
n-Octane	$CH_3(CH_2)_6CH_3$	-19

The octane rating of a particular petroleum distillate can be raised by blending in compounds of relatively high octane rating. Unfortunately, materials like benzene and toluene that tend to have high octane ratings also tend to be compounds with high value as industrial chemicals. As a consequence, these materials are more expensive than the low-octane fuel components.

A number of additives have been discovered that improve the octane rating of gasoline without using the expensive high-octane hydrocarbons. The most successful has been tetraethyl lead, commonly called *ethyl*.

On the average, there are about three grams of tetraethyl lead per gallon of gasoline. Other additives containing materials such as phosphorus, nickel, and boron also have been used, but have not become so popular as tetraethyl lead.

The "no-lead" fuels have no tetraethyl lead and obtain their high octane rating by other additives and by an increase in the high-cost, high-octane hydrocarbons. The presence of these higher-cost components is responsible for the increased cost of nonleaded fuels.

$$CH_3CH_2 \diagdown \quad \diagup CH_2CH_3$$
$$Pb$$
$$CH_3CH_2 \diagup \quad \diagdown CH_2CH_3$$

Tetraethyl lead

When first introduced, the no-lead fuels did not sell well, and most were subsequently taken off the market. They were replaced with low-lead fuels, which are a compromise between the two extremes and are lower in price than unleaded gasoline. With the advent of catalytic mufflers on cars, no-lead fuels are again being marketed because the lead in other gasolines will deactivate the catalysts in the mufflers.

combustion in the automobile engine

Although it is true that gasoline is a complex mixture of hydrocarbons, much can be learned about its characteristics in the gasoline engine by assuming that gasoline is a pure substance of formula C_8H_{18}. When allowed to burn in open air, this substance is converted to carbon dioxide and water.

$$2 \ C_8H_{18} + 25 \ O_2 \rightarrow 16 \ CO_2 + 18 \ H_2O$$

In air there is an excess of oxygen, and as a consequence the reaction proceeds completely as shown.

Unfortunately, there is usually insufficient oxygen in the internal combustion engine to give complete combustion of the gasoline to carbon dioxide and water. As a consequence, some gasoline passes through the engine completely unburned. Unburned gasoline is a principal source of hydrocarbons in the atmosphere. Most gasoline is burned, but not completely to carbon dioxide. In this case, carbon monoxide is produced instead of carbon dioxide.

$$2 \ C_8H_{18} + 17 \ O_2 \rightarrow 16 \ CO + 18 \ H_2O$$

If only carbon monoxide and water are produced, as indicated in the equation, only $^{17}/_{25}$ as much oxygen is required, and less energy is released from the gasoline. In an actual operating engine, some unburned hydrocarbons, some carbon dioxide, and some carbon monoxide are produced. Partial combustion of the hydrocarbons to alcohols, aldehydes, ketones, and carboxylic acids also occurs to some extent. The relative amounts depend on the condition and tuning of the engine. A recent analytical study of automobile exhaust has revealed the presence of nearly two hundred different hydrocarbons. Some of these were original constituents of gasoline, while others were formed in the automobile engine.

air pollution and the automobile

To achieve maximum energy from the fuel, the ideal carburetor setting for an automobile engine would allow air and gasoline to mix in the exact proportions required by the combustion equation. Unfortunately, with the present level of carburetor technology it is impossible to maintain this ideal setting accurately for a long period of time. Furthermore, many engines actually run better with a gasoline-to-air ratio that either contains too much gasoline (a rich mixture) or too little gasoline (a lean mixture).

Engines manufactured before the middle 1960s normally were designed to use a rich mixture. This led to much criticism of the automobile industry because of the emission of large quantities of unburned or partially burned hydrocarbons into the air and the consequent contributions to the smog problem. In response to this criticism, the automobile industry undertook a research program to correct the situation. As a result, the typical modern automobile has been redesigned to operate on a lean gas mixture. Hydrocarbon emissions from these redesigned engines are definitely lower than they were from earlier engines.

Unfortunately, chemical reactions other than combustion of hydrocarbons also take place in the automobile engine, and some of these are increased under lean conditions. The most important of these reactions is the interaction of nitrogen from the air with oxygen to form nitric oxide.

$$N_2 + O_2 \rightarrow 2\,NO$$

This reaction is favored by the high temperature in the combustion chamber of the engine and by the presence in the cylinder walls of transition metals that can act as catalysts. Currently, the internal combustion engine is believed to be the principal source of nitric oxide in the atmosphere of large cities such as Los Angeles and Chicago. The effects of nitric oxide and its further reactions have been discussed already.

Lead from tetraethyl lead is also a source of air pollution. There is evidence of high lead concentrations in ocean water and rainfall in areas surrounding major cities. A recent study has indicated that the lead concentration in snow in Greenland has been increasing steadily since the beginning of the Industrial Revolution and has increased dramatically since 1940, when the automobile became important (see Figure 17-2). The validity of this study has been questioned, however, because of the heavy use of Greenland for military aircraft since 1940.

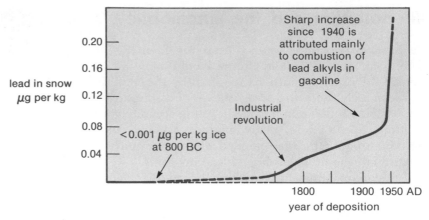

Figure 17-2. Lead content in snow in Greenland (Copyright *Chemistry in Britain*, 1971, 7, 55)

It is known that lead tends to concentrate in bone tissue. It is also known that lead compounds in large quantities can cause serious health problems and even death. The health hazard created by low levels of lead is still speculative, however. Nevertheless, removal of lead is undoubtedly a good idea, even if no major harmful effects can be attributed to it directly.

Unfortunately, removal of lead from gasoline may provoke other problems that do not exist now. The high-octane compounds that are added to replace lead in gasoline to improve the octane rating are usually aromatic compounds such as benzene and toluene. Under the conditions existing in the automobile engine, these compounds tend to condense to form polycyclic aromatic hydrocarbons. Many of these have been shown to be carcinogenic. In other words, they can cause cancer. Thus by eliminating the danger from lead, the danger of cancer may be increased by the presence of these new products in the air.

elimination of automobile pollution

The elimination of pollution by the automobile has evolved into both a technical and a political problem. Emotions run high whenever the subject is discussed. Some people maintain that the problems inherent in the internal combustion engine are so complex that they can never be solved. These people maintain that a totally new form of propulsion must be devised. Unfortunately, no acceptable new form of propulsion seems to be imminent. The Wankel (rotary) engine may offer a major improvement, but probably is not the ultimate answer.

Above left: Automobile emissions can be analyzed by equipment such as this. Tests reveal how well an automobile is functioning and burning its fuel. *(Courtesy Environmental Protection Agency)*

Above right: Sundancer I, a recently developed experimental electrical car, uses the standard lead storage battery as a power source. Electrically powered cars, trucks, and buses are still largely in the experimental stage, however. *(Courtesy Electric Vehicle Division, E.S B., Inc.)*

Left: A see-through model of a Wankel rotary engine. *(Courtesy Edmund Scientific Co., Barrington, N.J.)*

Other people claim that the problem is political. They maintain that if tough new laws were passed, the automobile industry would be forced to make the required improvements.

As with most emotional issues, reality probably lies between the extremes. Automobile manufacturers were slow in facing the problems that their product created, but they seem to be responding at present. To the extent that the technology to solve the problem already exists or can be quickly developed, more stringent laws would encourage its application. To the extent that the technology does not exist or cannot be developed rapidly, new laws would have no effect and would be unenforceable.

The real questions that must be answered are: what can be done now to solve the problem of automobile pollution, and what can be reasonably expected in the near future? These questions are not easy to answer with any accuracy because adequate information is not yet available. The reason for inadequate data is that work in this area has only begun relatively recently.

The problems involved in studies of automobile pollution are complex. Not only is it difficult to make many of the analytical measurements, but also the final data is frequently subject to more than one interpretation. The Greenland lead study is a good example of the problems that may arise in interpreting data. When several interpretations are possible, more data must be collected to resolve the differences. This is the normal way scientific knowledge is acquired and is an example of the scientific method in operation. Even the differences of opinion are part of the process. Unfortunately, the general public frequently is frustrated and annoyed by the process. Resolution of differences is frequently slow, especially if the problems are complex. Meanwhile opponents become quite vocal and people do not know whom to believe. The whole environmental issue is still in this stage of development.

Some companies have developed catalytic converters that are being installed in the exhaust systems of automobiles. The companies claim that these devices will essentially eliminate hydrocarbon and nitrogen oxide emissions from the automobile. Furthermore, it has been claimed that gasoline mileage is increased when cars are equipped with catalytic converters because other types of pollution-control devices can be removed. Unfortunately, the catalysts in these converters tend to be deactivated by lead compounds. This is one of the main reasons for the effort to eliminate tetraethyl lead from gasoline.

carbon dioxide as a pollutant
—the greenhouse effect

Normally, carbon dioxide is not considered a pollutant by most people. It is an important constituent of the air and is absolutely necessary for life. Nevertheless, some people are becoming alarmed by the large quantities of carbon dioxide that enter the atmosphere annually. Not only are the normal surface sources of carbon such as wood being consumed at a high rate, but also underground sources such as coal and oil are being consumed at a tremendous rate. Plant life, through the process of photosynthesis, uses carbon dioxide from the atmosphere, but it cannot keep up with the present output of carbon dioxide from combustion. The situation is further aggravated by the fact that as cities expand, more of the earth's surface is being covered by buildings and paving materials. Consequently, the carbon dioxide concentration in the air has been slowly increasing.

In 1890, the carbon dioxide concentration in the atmosphere was estimated to be about 290 ppm. In the early 1960s, it was estimated to be about 315 ppm. The yearly rate of increase in the early 1960s was estimated to be 0.7 ppm. The increase has been attributed primarily to consumption of the fossil fuels, petroleum and coal.

It has been suggested that this increase in the carbon dioxide content of the air will raise the temperature of the earth. This could happen because light energy from the sun tends to be of short wavelengths that are not absorbed by the atmosphere (including carbon dioxide). After interaction with the earth's surface, however, much of this energy is converted to thermal or infrared energy. Infrared energy is absorbed by carbon dioxide and thus is not permitted to escape from the earth. As the concentration of carbon dioxide in the air is increased, the efficiency of the energy-trapping process is increased, thus causing a rise in the earth's temperature. This phenomenon is called the *greenhouse effect*.

It has been estimated that the temperature of the earth would increase about 2–4°C, if the carbon dioxide content of the air were doubled. Furthermore, it would be possible to quadruple the present carbon dioxide content if all known quantities of fossil fuels were to be consumed within the next 500 years, leading to a temperature increase of about 7–12°C.

An increase in temperature of this magnitude could have disastrous effects. Plants and animals would have to seek new territories or become extinct. The polar ice caps would melt, causing a rise in the ocean level and the submergence of major coastal cities.

There is almost no end to the potential doom that has been forecast by proponents of the greenhouse effect. Actually, there is no real evidence to support or deny the existence of such an effect, and the accuracy of the figures given here is subject to question. Furthermore, the earth seems to have been cooling during the past few years. At first, this would appear to disprove the greenhouse theory. Its proponents maintain, however, that this is just coincidental, because the increase in carbon dioxide content has been balanced by an increase in the particulate matter content of the air. Particulate matter tends to reflect the heat from the sun, thus counterbalancing the greenhouse effect. Proponents of the greenhouse effect argue that as we learn to control particulate emissions, the present equilibrium will be disturbed, and the greenhouse effect will be apparent. At this time, no one knows for sure what to expect.

references

Alexander, Tom, *Some Burning Questions about Combustion*, Fortune, **81** (2), 124 (1970).

Bryce-Smith, D., *Lead Pollution—A Growing Hazard to Public Health*, Chemistry in Britain, **7** (1), 54 (1971).

Bryce-Smith, D., *Lead Pollution from Petrol*, Chemistry in Britain, **7** (7), 284 (1971).

Bylinsky, Gene, *The Long, Littered Path to Clean Air and Water*, Fortune, **82** (4), 112 (1970).

Davenport, John, *Industry Starts the Big Clean Up*, Fortune, **81** (2), 114 (1970).

Demaree, Allan T., *Cars and Cities on a Collision Course*, Fortune, **81** (2), 124 (1970).

Greek, Bruce F., *Gasoline-Antipollution Forces Bring Marked Changes to Petroleum Refining Industry*, Chemical and Engineering News, **48** (24), 38 (1970).

Hoar, P. P., *Our Environment 3. The Problem of Corrosion*, Chemistry in Britain, **6** (5), 204 (1970).

Mahler, E. A. J., *Our Environment 2. Air Pollution*, Chemistry in Britain, **6** (5), 201 (1970).

Maxwell, Kenneth E., Editor, *Chemicals and Life*, Dickenson Publishing Company, Inc., Encino, Calif., 1970.

Mills, A. L., *Lead in the Environment*, Chemistry in Britain, **7**, 160 (1971).

O'Sullivan, Dermot A., *Air Pollution*, Chemical and Engineering News, **48** (24), 38 (1970).

Rose, Sanford, *The Economics of Environmental Quality*, Fortune, **81** (2), 120 (1970).

Ways, Max, *How to Think about the Environment*, Fortune, **81** (2), 98 (1970).

Cleaning Our Environment—The Chemical Basis for Action, American Chemical Society, Washington, 1969.

study questions

1 Would an atmosphere containing an excess amount of oxygen be considered polluted according to the definition of pollution used in this chapter?

2 Is air pollution something that has been known for a long time?

3 Approximately what portion of the total world air pollution is caused by man?

4 What is the source of ozone in a smoggy city? Can you name another natural source?

5 What are the undesirable effects of ozone in the atmosphere?

6 What are some of the major sources of sulfur dioxide in the air?

7 Why is sulfur dioxide considered a particularly foul pollutant?

8 Describe a process that is being considered for removing sulfur dioxide from exhaust stacks of electrical power plants.

9 How is electricity generated in your area? Has the power plant produced any pollution of the environment?

10 Coal gasification appears to be promising as a method of producing synthetic natural gas. What problems might this process create if it were to become a major energy-producing process?

11 What is the function of tetraethyl lead in gasoline?

12 What are the reasons for wanting to remove lead from gasoline, and what are the difficulties associated with this process?

13 What are some pollutants produced or emitted by the automobile?

14 What is the greenhouse effect?

15 One area of concern in the pollution problem is differing opinions of the extent of the problem. What is the solution to this?

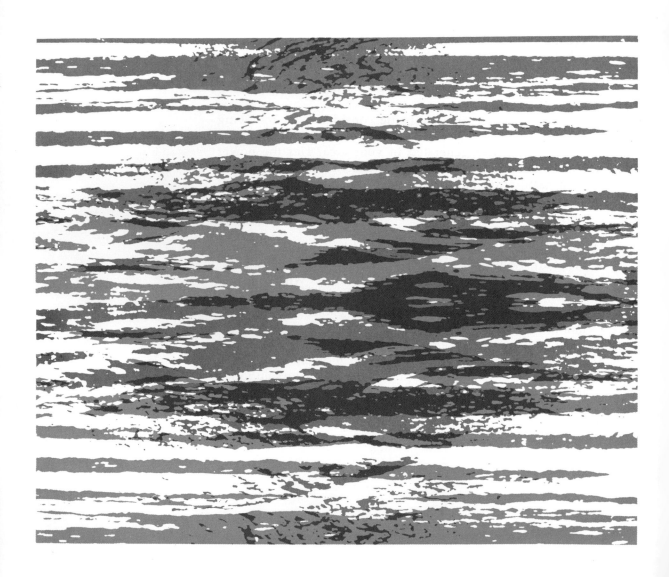

chemistry of water quality and solid waste

Pollution is not solely a problem of the atmosphere. Currently, much of the earth's surface is faced with problems of water pollution as well. In addition, pesticides and even household trash can pose pollution problems. These situations, as with air pollution, are largely chemical in nature and require chemical solutions.

alkylbenzene sulfonate (ABS)
eutrophication
nitrilotriacetic acid (NTA)
methyl mercury
pesticide
BOD test
primary sewage treatment
secondary sewage treatment
sanitary landfill

water pollution

Water pollution has been a problem for hundreds of years The Thames River, which flows through London, has been badly polluted since at least the eighteenth century and has long been a target for cartoonists. The pollution was so bad 50 years ago that for a time there were no fish in the river.

Today many lakes and rivers in the United States have become so badly polluted that they are no longer safe as sources of drinking water without extensive purification procedures. Often the required purification processes are unavailable, and drinking-water supplies in these areas have declined in quality to dangerous levels. Lake Erie has become so badly polluted that many people have called it dead. Much of the lake's southern shore has been declared unfit for swimming. The Cuyahoga River, which flows into Lake Erie through Cleveland, Ohio, has become so badly polluted that for years people jokingly called it a fire hazard. The jokes ceased in 1969, however, when a fire did break out on the river and severely damaged a railroad bridge. Even the oceans, long thought to have an infinite capacity for refuse, are becoming intolerably polluted along many coastal areas. In spite of this, the oceans still are used as depositories of raw sewage and garbage.

As the world population has increased, the pollution problem also has increased. More sewage as well as more industrial and agricultural wastes have been generated. Furthermore, as society has become more complex, wastes have become more complex. Pesticides, hydrocarbons, detergents, phosphates, nitrates, and many other materials are being produced and used at a rapidly increasing rate. Most of these materials ultimately find their way into the waterways of the country as pollutants. As a consequence of the pressure of the expanding needs of society, the quality of the waterways is deteriorating at an accelerating pace.

The problem has reached beyond the borders of individual countries and is worldwide in scope. Rivers in Europe sometimes flow through several countries, becoming dirtier as they flow toward the sea. DDT, the insecticide that has brought malaria largely under control, has been found all over the world, even in remote places like Antarctica where it has never been used.

The situation is not hopeless, however. The Thames River, for example, is now cleaner than it has been for years, and several species of fish now live again in its waters. This revitalization has been accomplished through a concerted effort to prevent raw sewage and industrial effluents from entering the river or at least to reduce the amount. The result is a long way from perfection, but does demonstrate that the trend toward pollution can be reversed.

the detergent problem

Several years ago sewage-treatment facilities in some areas became almost inoperative because of the massive amount of detergent foam produced in their aeration tanks. The problem soon spread, and some lakes and rivers were observed to have "soap suds" floating on their surfaces. Even kitchen faucets in some parts of the country were dispensing foam because of residual detergent in the water.

This was the first major, detergent-caused, pollution problem to receive national attention. Both the detergent industry, which was directly involved, and the petroleum industry, which was involved as a supplier to the detergent industry, spent several years and millions of dollars trying to solve the problem. Finally it was discovered that the source of the problem was alkylbenzene sulfonate (ABS), the most common surfactant (active ingredient) then in detergents. (See Chapter 15 for a further discussion of soaps and detergents.) ABS (II opposite) was made by reacting a propylene tetramer (I opposite) with benzene, followed by sulfonation.

ABS was an excellent detergent material, and it could be made easily from inexpensive petroleum hydrocarbons and sold at low

Occurrence of detergent foam in rivers and lakes led to the development of biodegradable detergents. (*UPI Photo*)

cost. Unfortunately, ABS made from propylene is quite stable. In nature, bacteria can degrade it to simple compounds only with difficulty. Thus ABS began to build up in the groundwater and waterways of the country.

Living organisms have a highly developed ability to degrade linear hydrocarbon chains by the process of β oxidation. (β oxidation is discussed in Chapter 13.) They have a more limited ability to degrade branched hydrocarbon chains. The branched methyl groups on the ABS molecule were found to be inhibiting bacterial action on ABS.

The solution to the pollution problem required a change from propylene to ethylene as the feedstock for the surfactant manufacturing process.

$$CH_2{=}CH_2 + CH_2{=}CH_2 + CH_2{=}CH_2 + CH_2{=}CH_2 + CH_2{=}CH_2$$

Ethylene

$$\downarrow$$

$$CH_3CH_2CH_2CH_2CH_2CH_2CH_2CH_2CH{=}CH_2$$

$$\downarrow$$

$$\downarrow \text{(several steps)}$$

$$\downarrow$$

$$CH_3CH_2CH_2CH_2CH_2CH_2CH_2CH_2CH_2CH_2{-}\!\!\!\left\langle\bigcirc\right\rangle\!\!\!{-}SO_3^-Na^+$$

The new ABS molecule is structurally similar to the original molecule and is also a good detergent. An alternate source of the linear hydrocarbons needed for this process is a fraction of cracked petroleum.

The commercial changeover to linear ABS detergents was made between 1963 and 1965. It was essentially complete by the end of 1965. Since then, soap suds gradually have disappeared from waterways and are no longer a popular subject of complaint.

A new detergent problem has arisen, however. This is the phosphate problem. As discussed in Chapter 15, detergents contain large quantities of phosphates in addition to surfactants. It has been estimated that two billion pounds of phosphates are used per year by the detergent industry. Detergents represent about 70% of the sewage phosphate in the United States. The balance comes from human waste and agricultural and industrial effluents. Typical municipal sewage contains about 10 ppm of phosphate.

Widespread use of phosphates has been linked to the development of serious problems in the world's waterways. The problem is termed eutrophication. Eutrophication is caused by excess nutrients in the water, enabling algae to grow in great abundance. Huge algal blooms can be seen in lakes and rivers where sufficient nutrients

Fishkills of this type are caused by eutrophication and pollution of water. (*Courtesy Environmental Protection Agency*)

are present. When algae die, they sink to the bottom of the lake or river and begin to decay. The decay process consumes oxygen and soon depletes the oxygen content of the water, killing fish and other aquatic life. Once the oxygen is gone, anaerobic bacteria (bacteria capable of living in the absence of oxygen) begin to feed on the decaying algae. The life processes of anaerobic bacteria produce compounds like hydrogen sulfide, which cause the foul, putrid odor associated with decaying organic matter. Thus the water develops foul smell and taste and becomes generally objectionable. Many of the problems in Lake Erie have been associated with eutrophication.

The major nutrients required by algae are phosphates and nitrates. Normally, there is such a small amount of phosphate in the water that the phosphate content is believed to limit the amount of algae that can grow. Although some disagreement exists, most scientists concerned with the problem believe that there is normally plenty of nitrate present in the water and that nitrate cannot limit the algal population. Furthermore, some blue-green algae can use nitrogen from the air and do not need nitrate. Because of the belief that phosphates stimulate algal growth, there has been an effort to eliminate or reduce the phosphate content in detergents. In areas with soft water, most, if not all, phosphates could be excluded from detergents with no ill effects. This regional difference would require manufacturers to market different detergents in different regions of the country. Detergent manufacturers have been reluctant to do this because of the increased cost.

Another possibility that has been considered involves the production of several grades of detergent. Thus there might be a low-phosphate detergent for only slightly soiled clothes and a heavy-duty, high-phosphate detergent for extremely greasy clothes. Again, there would be increased cost because of more complicated manufacturing, packaging, and marketing operations. Furthermore, this suggestion would require educating people about the proper detergent for each kind of wash, another problem that the detergent industry has been reluctant to tackle.

A more satisfactory solution would be to find a substitute for phosphate. So far this has proved a difficult task. Phosphates are inexpensive, relatively nontoxic, harmless to clothes and washing equipment, and readily available. Generally, substitute compounds have proved expensive, toxic, or harmful to clothes and washing equipment.

For several years, nitrilotriacetic acid, commonly called NTA, was thought to be a suitable substitute for phosphate.

NTA proved to be economical, versatile, effective, and relatively nontoxic. Furthermore, it could be biodegraded to harmless materials. NTA looked so promising as a phosphate substitute that it was adopted in Sweden, and two major chemical companies in the United States began building multimillion dollar plants to manufacture it. In December of 1970, however, the federal government banned the use of NTA in the United States because of data indicating harmful effects of NTA on rat embryos.

Several other materials have been considered, but have not been accepted because of high cost or other disadvantages. One such material is sodium citrate.

Sodium citrate would appear to be an ideal substance in many respects. It is a weak acid and can act as a buffer. It has three acid groups and can act as a water-softening agent. It is a constituent of biological processes (the tricarboxylic acid cycle) and is biodegradable and should be nontoxic. Unfortunately, it is difficult to synthesize on a commercial scale and must be obtained from natural sources.

For a while, a few detergents were advertised as being phosphate-free. Unfortunately, most of these contained strong bases (alkalis) that could cause chemical burns and also could damage some metal parts of washing machines. These detergents now are required to carry warnings on their labels indicating this danger. Most have been taken off the market. Presently there does not

$$N \underset{\displaystyle CH_2COOH}{\overset{\displaystyle CH_2COOH}{\big< }} CH_2COOH$$

NTA

$$\begin{array}{c} CH_2 - COO^-Na^+ \\ | \\ HO - C - COO^-Na^+ \\ | \\ CH_2 - COO^-Na^+ \end{array}$$

Sodium citrate

appear to be a suitable phosphate substitute in the offing. Since September 1971, the federal government has recommended that phosphate detergents be used in preference to these new substitutes, as the least objectionable material at this time.

A final possible solution to the problem involves removing the phosphate from the water. This approach will be discussed along with other sewage-treatment problems in a later section.

Like most problems of pollution control, the phosphate problem is important. To date, solutions have been only partially successful. Although it is important that the problem be solved as soon as possible, it is equally important that the solutions not create more problems than the one they were designed to cure.

metals in the water

The problem of metal poisoning from water first came to the attention of the American public in 1970, with the discovery of high mercury levels in fish from Lake Erie. The source of the mercury was traced to industrial plants in adjacent states, primarily Michigan, which produced chlorine by the mercury-cell process. The economics of this process have been affected drastically by the previously unsuspected need to prevent mercury's escaping from these plants to the environment. Consequently, some of these plants have been shut down.

Prior to the 1970 discovery of high levels of mercury in fish, it was believed that the small amount of mercury escaping from these plants would not be a health hazard. Mercury metal is relatively inert and quite dense, so it was assumed that mercury settled to the bottom of the river or lake and stayed there. Since 1970, however, it has been determined that certain anaerobic organisms in the water can convert mercury into an organometallic compound known as methyl mercury.

$$CH_3—Hg—CH_3$$

Methyl mercury

Methyl mercury is soluble in water and in organic tissue. Once it is released into the water by the synthesizing organism, methyl mercury slowly finds its way into the food chain and becomes concentrated in fish tissue.

Since its original discovery in Lake Erie fish, mercury has been found in fish in many parts of the country. In some places, like the Rocky Mountains, there is no obvious relationship to industrial activity. In this case, the source of mercury must be natural deposits. Even the oceans are not free of the contaminant. Excessive concentrations of mercury have been found occasionally in tuna and routinely in swordfish. Mercury contamination in swordfish is so consistent that the government has recommended that people in the United States no longer eat it, essentially shutting down an entire industry.

In 1953, many people in Japan died or became ill after eating mercury-contaminated fish. The cause of these deaths was not discovered until several years later because mercury was difficult to detect in small quantities until relatively recently. This was the first recorded instance of the consequence of mercury pollution, however. In recent years new techniques like atomic-absorption spectroscopy have revolutionized the analysis of many kinds of trace metals. These new developments in analytical techniques have proved valuable in pollution studies because mercury is not the only metal prompting concern.

The problem of lead in the air has been discussed already. Lead also finds its way into waterways. Lead concentration in the ocean near Los Angeles, for example, is far in excess of that in other areas. Other metals also are found in water and air as a consequence of industrial and refining operations. Some examples are copper, zinc, cadmium, beryllium, nickel, and vanadium. Since these metals have been found to cause liver and kidney ailments, high blood pressure, mental disorders, and other health problems, their detection by proper analytical methods is important.

A major problem in metal-pollution control is determining acceptable, safe limits. Presently insufficient data are available to establish realistic limits on the amount of these materials that can be tolerated safely by living organisms. A maximum acceptable mercury concentration of 0.5 ppm in fish has been established somewhat arbitrarily by the Food and Drug Administration. This is about 1% of the average mercury concentration found in the Japanese fish that caused mental disorders and death. No one knows, however, if this limit is appropriate or not. Perhaps higher levels could be tolerated, at least for occasional exposure. On the other hand, continuous exposure at even this level might prove harmful over a long period. It is known that some metals do accumulate in the body over a long period of low-level exposure. Furthermore, metals, unlike organic impurities, are not degradable.

The situation is complicated further because the body requires some of these metals, such as manganese, in small amounts as metabolic regulators. Thus complete elimination of these metals from the food supply could be just as disastrous as having too much present. A whole new area for research seems to be opening as a consequence of these problems.

pesticides

Pesticides are chemicals used to kill insects, fungi, weeds, and other pests. Thus, the term *pesticides* includes insecticides, fungicides, herbicides, and similar materials. Use of these compounds has increased remarkably in the past 30 years and has resulted in expanded crop yields, disease control, and other benefits to mankind.

Unfortunately, there are harmful effects of pesticides, some of which are just now becoming apparent. Like nonbiodegradable detergents described earlier, many pesticides have a long lifetime in the environment. Chlorinated pesticides, such as DDT, appear to be the worst offenders of this type. Part of the reason for this may be that chlorinated pesticides have been the most heavily used and thoroughly studied. Figure 18-1 shows some common pesticides and their chemical structures.

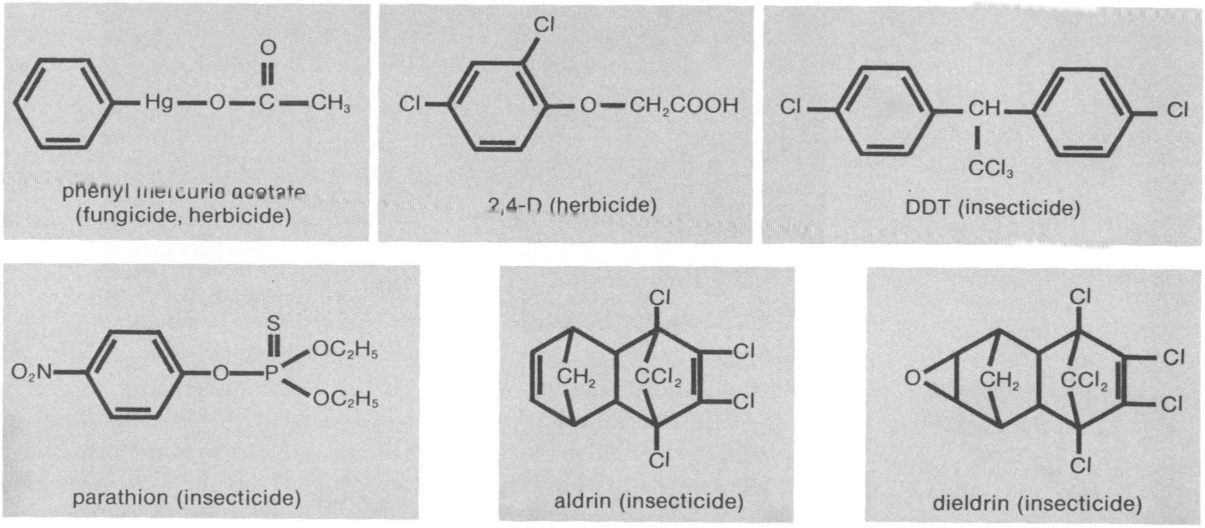

Figure 18-1. Some common pesticides

Near areas of heavy pesticide use, such as agricultural areas of the United States, high concentrations of DDT, toxaphene, and dieldrin have been found in the fatty tissue of wildlife. Even in remote areas like Antarctica, DDT residues have been found in penguins, seals, and other animals that feed on marine life.

The effect of low levels of pesticides on animal and plant life has been only partially studied. Nevertheless, some facts are emerging.

DDT appears to cause the formation of bird eggs with thin shells, which may lead to extinction of some bird species. (*From David Johnston, John Netterville, James Wood, and Mark Jones, Chemistry and the Environment. Published by W. B. Saunders Company, Philadelphia, 1973*)

The population of some wild birds has been decreasing in recent years. This decrease has been correlated with the increased use of chlorinated pesticides since World War II. Deaths in some birds have at times been attributed directly to high DDT levels in their tissues. DDT also has been reported to lead to the production of thin eggshells by some bird species. Survival rates are low in species producing thin eggshells.

To reduce the harmful effects of pesticides on the environment, some substances, such as DDT, have been banned in the United

One alternate approach to insect control is the use of pheromones. These naturally occurring chemicals are frequently secreted by insects of one sex to attract the opposite sex. Synthetically produced pheromones may be used to attract insects to a poison. This technique makes it possible to eliminate insects selectively, without spreading pesticide over large areas.

States. Elimination of DDT from the market has led to a search for acceptable alternatives. Use of natural enemies, such as birds, bats, and even viruses, is being studied as an alternative to massive pesticide application. In most cases, these new approaches are still in an experimental stage, and their effectiveness has yet to be demonstrated.

sewage treatment

For centuries, the main device for waste disposal was the privy or outhouse. In recent times, in the United States and other developed countries, the septic tank has become widely used. In sparsely settled areas and when dealing only with relatively simple organic wastes, these two methods of waste disposal are quite adequate, because land and water have a remarkable ability to purify themselves.

In more heavily populated areas, however, the quantity of waste material may exceed the ability of the natural processes to purify it. Furthermore, the accompanying industrial growth creates more complex waste problems than natural systems can handle. Thus cities usually have a sewer system to handle waste. Unfortunately, the sewer systems in most cities are inadequate to meet the needs of the cities, and a large number of septic tanks are still in use. Furthermore, many cities use their sewer systems merely for collection and then discharge the material, untreated, directly into the nearest body of water. Even cities with treatment facilities frequently find that these facilities are inadequate to meet the demand and so still discharge large quantities of raw sewage directly into rivers, lakes, or the ocean.

The major problem associated with the degradation in water of raw sewage is its requirement for oxygen. As the sewage is decomposed by various bacteria and other organisms, oxygen is consumed from the water. If the amount of sewage is large enough, all the oxygen in the water will be consumed, and fish will die.

To measure the extent to which water has been affected by the presence of sewage, a test known as the BOD test has been developed. This test takes five days to complete and is a measure of the biochemical oxygen demand (BOD) of the water. The test has been accepted as the standard test of water quality and is used to test the water leaving sewage-treatment plants. There are now legal limits in most areas concerning the allowable BOD that a sewage-plant discharge cannot exceed.

Water purity can be measured by commercially available testing kits. The kit shown is for determining hydrogen sulfide in water. (*Courtesy Edmund Scientific Co., Barrington, N.J.*)

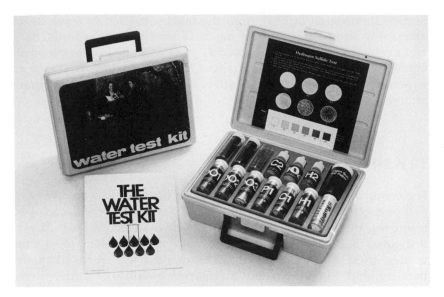

The BOD test has some disadvantages, since it takes five days to complete. For this reason, attempts have been made to find a faster test. Two other tests have been developed and are known as the *COD* (chemical oxygen demand) and the *TOD* (total oxygen demand) *tests.* Unfortunately, the results of these tests do not necessarily correspond with those of the BOD test, since the COD and TOD tests depend on the nature of the pollutants present. Therefore these tests have not replaced the BOD test.

Sewage treatment generally is carried out in one of three stages known as *primary*, *secondary*, and *tertiary* treatments. (See Fig. 18-2.) Most major cities now have primary and secondary treatment facilities, and the federal government is beginning to aid in constructing such facilities in cities where they do not exist presently. Few places have tertiary treatment facilities, however.

Primary sewage treatment is mechanical. Solid objects like sticks and rags are separated by a coarse screen. The liquid material that passes through the screen flows into large settling tanks, where insoluble material settles and forms a sludge. This process removes about one-third of the pollutants from the water.

Secondary sewage treatment is largely biological. In older sewage plants, the effluent from the primary facility filters slowly through a bed of rocks. During this process, bacteria consume most of the organic material. Newer plants bubble air through the tank to accelerate the growth of bacteria and the subsequent decomposition of organic material. A more recent experimental process has been developed that employs pure oxygen for the activation operation.

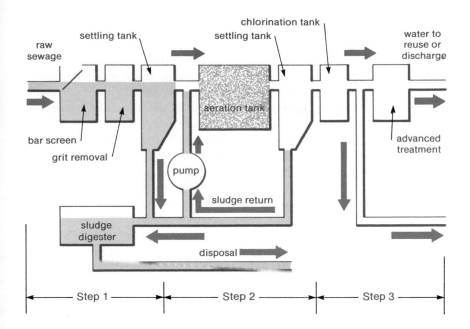

chlorination tank

settling tank

settling tank

water to reuse or discharge

raw sewage

aeration tank

bar screen

advanced treatment

grit removal

pump

sludge return

sludge digester

disposal

Step 1   Step 2   Step 3

Step 1: Primary treatment

Removes approximately 40% BOD. ■ Bar screen catches large objects; others go through grinder. ■ Water flows slowly through grit chamber, sand and gravel settle. Can be used for landfill as it contains little organic matter. ■ Water stands in settling tank. ■ No direct reuse of water possible.

Step 2: Secondary treatment

Can remove 45-55% BOD (total 85-95% BOD removal). ■ Aeration helps growth of aerobic organisms (activated sludge process). ■ Trickling filters (bed of coarse stones) may replace aeration; organisms on stones feed on organic material. ■ Sludge used as landfill or fertilizer; probably high in metallic content. ■ Water can be used for irrigation, recreation, industrial supply. Water will still contain dissolved nutrients, some types of organic materials.

Step 3: Tertiary treatment

Most expensive treatment. ■ Used to remove organic chemicals, nutrients, and excessive salts. ■ May use chemical coagulation, distillation, reverse osmosis, etc. ■ Can use water for all uses, including irrigation, industrial supply, groundwater recharge, drinking (if meets Public Health Standards).

Figure 18-2.   Treatment of Sewage

Secondary treatment removes about 90% of the biodegradable materials in the water. Most municipalities chlorinate the effluent from this process to kill disease germs and then discharge the effluent directly into the nearest river, lake, or ocean.

Unfortunately, as the way of life becomes more complex, pollutants in the water also become more complex, and the need for more sophisticated techniques of purification increases. Pesticides, detergents, nitrates, metals, and other contaminating materials also make their way into sewer systems. Most of these materials are not removed by primary and secondary sewage-treatment facilities and are dumped into the waterways. Cities downstream from these plants often use these waterways as part of their water supplies. Furthermore, nitrates and phosphates serve as nutrients for algae and lead to eutrophication. Thus, even with primary and secondary treatment, the waterways still become polluted.

To prevent the problems associated with these contaminants in waterways, tertiary treatment facilities are now being considered. Several purification schemes have been devised, but most are still in the experimental stage. Typical tertiary treatment plants use materials such as activated charcoal and diatomaceous earth to absorb the remaining pollutants.

Frequently the water is chlorinated to kill any remaining bacteria or viruses before discharge into the waterways. One such process is used at Lake Tahoe, on the California-Nevada border. Effluent from the Tahoe sewage-treatment facility discharges into a large man-made lake and is pure enough for recreational and irrigational purposes.

solid-waste disposal

Solid-waste disposal methods have not reached the degree of sophistication found in disposal methods for air and water waste. The sheer bulk of solid material being generated is making this an area for concern. It has been estimated that 10 pounds of solid waste are generated per person per day in the United States. Of this material, a little over 5 pounds is collected by various agencies. As the economy continues to grow, the amount of solid waste generated per person also grows. Furthermore, the population itself is still growing, which magnifies the problem even more.

Municipal waste is a mixture of almost every type of material used by the general population. It contains paper, plastics, metals, glass, wood, and many other materials. Many of these materials could be reclaimed and thus reused if an economical separation procedure could be developed. Glass, metals, and some plastics easily fall into this category. Several schemes have been tried on an experimental basis for separating the components for recycling, but none has achieved widespread use. Some of these methods probably will be commercialized in the near future, however, as more people become aware of the problem and perfect the required equipment.

At present, incineration and sanitary landfill are the preferred methods for solid-waste disposal. Home incinerators are largely a

The sanitary landfill is a major way to dispose of solid waste. Refuse is piled in a large pit or natural gorge and covered with dirt. Eventually the land is reclaimed for recreational and other uses. (*Photographs by Treva Richards*)

thing of the past in the modern United States city, but they still are used in rural areas. Municipal incinerators commonly are used, however, and are undergoing rapid development to meet air-pollution standards. Today municipal incinerators must meet rigid regulations concerning the amount of pollutants that may be emitted. To achieve the required efficiency, the refuse usually is burned with excess air in a multistage incinerator. This ensures that all volatile material is burned to simple oxidation products like carbon dioxide and water. Usually, sulfur dioxide and nitrogen oxide production is assumed to be small compared with other sources, and is ignored. More rigid regulations in the future may force control of even these substances. Scrubbers or electrostatic precipitators are being installed to prevent the emission of solid fly ash.

A growing problem in incinerators is the formation of hydrogen chloride. Hydrogen chloride is produced primarily by the combustion of chlorinated plastics. Hydrogen chloride can damage metal parts of incinerators and become a harmful air pollutant if allowed to enter the atmosphere. Polyvinylchloride (PVC) is a major contributor, since it contains about 50% chlorine by weight. The amount of chlorine-containing plastics being discarded is small at present, but use of these plastics in packaging and other consumable products is increasing.

In areas like Los Angeles, where air pollution is already a major problem, sanitary landfill is the preferred method of solid-waste disposal. In this process, refuse is spread in thin layers and compacted with bulldozers. After a 10-foot layer of refuse has been deposited, it is covered with a thin layer of clean soil. A thin layer of soil also is added at the end of each working day, to prevent wind from blowing the material around and to maintain sanitary conditions. When the area has been filled completely, it is covered with a final layer of soil two to three feet thick.

Decomposition of the organic materials in the refuse is accomplished by anaerobic bacteria. Methane and carbon dioxide are produced by these bacteria. These gases slowly escape through the surface. The rate of gas evolution is sufficiently slow, however, that the reclaimed land can be used for recreational purposes such as parks and golf courses. A ski slope is planned for the Chicago area, to be built from a sanitary landfill shaped like a small mountain. The problem of methane accumulation and the settling of the land makes the construction of buildings on or near a sanitary landfill inadvisable.

Potential water pollution from a sanitary landfill must be considered when selecting a site. This is especially true in areas with heavy rainfall that can dissolve the carbon dioxide and other decomposition products in the refuse. If these substances are leached into groundwater, they increase the hardness of the water and cause other potential problems. Bacterial contamination also may be a significant concern.

Although sanitary landfill probably will continue as the preferred method of solid-waste disposal for several years, it cannot be the ultimate answer. There is a limit to the available places suitable for landfill operations and close enough to large cities to be economical. Eventually these will be filled. When this happens, or even before, alternate methods must be adopted.

references

Bylinsky, Gene, *The Limited War on Water Pollution*, Fortune, **81** (2), 103 (1970).

Bylinsky, Gene, *The Long Littered Path to Clean Air and Water*, Fortune, **82** (4), 112 (1970).

Bylinsky, Gene, *Metallic Menaces to the Environment*, Fortune, **83** (1), 110 (1971).

Davenport, John, *Industry Starts the Big Clean Up*, Fortune, **81** (2), 114 (1970).

Epstein, Samuel S., *NTA*, Environment, **12** (7), 3 (1970).

Mellanby, Kenneth, *Our Environment 1. Environmental Pollution*, Chemistry in Britain, **6** (5), 198 (1970).

Rose, Sanford, *The Economics of Environmental Quality*, Fortune, **81** (2), 120 (1970).

Ways, Max, *How to Think about the Environment*, Fortune, **81** (2), 98 (1970).

Biodegradability of Detergents—A Story about Surfactants, Chemical and Engineering News, **41** (11), 102 (1963).

Union Carbide Will Make Detergent Alkylate, Chemical and Engineering News, **41** (24), 25 (1963).

New Technology Is Resulting from Syndet Raw Materials Switch, Chemical and Engineering News, **41** (44), 138 (1963).

Switch to Soft Detergents Set for June, Chemical and Engineering News, **43** (22), 26 (1965).

study questions

1 Is water pollution worse today than 50 years ago in all parts of the world? Explain.
2 What substance was responsible for producing "soapy" drinking water in the early 1960s? How was the problem solved?
3 How would removing phosphate ions from the waterways reduce eutrophication in lakes and streams?
4 Several substitutes for detergent phosphates have been considered. Why have these not proved satisfactory?
5 How do you think the phosphate problem should be solved?
6 Why are metals such as mercury a problem in the waterways?
7 The presence of small amounts of heavy metals such as mercury, lead, and cadmium in the environment can be quite dangerous. Why has this form of pollution only recently been recognized?
8 What problems exist in metal-pollution control?
9 What are some beneficial effects derived from the use of DDT?
10 What are some harmful effects derived from the use of DDT?
11 What matters of concern exist regarding pesticide use and control?
12 What is one major problem with the BOD test?
13 Most city sewage facilities are equipped to perform primary and secondary treatments on sewage. What does each of these processes remove from raw sewage?
14 Why are tertiary treatment facilities in sewage facilities desirable for further processing of sewage?
15 What are some alternatives for solid-waste disposal? Which of these are used in your hometown or locally? What do you think will be used in your town 25 years from now?

1 Nitrate ion, carried into the waterways by excessive use of nitrate fertilizers or cattle feedlots, can cause the death rate of infants to increase. At the same time, the high yield of food from heavily fertilized farms is necessary for the farmer to make a living and for the world population to be adequately fed. What would you suggest as a possible solution to this dilemma?

2 How would you locate the source of the problem if your water supply were found to be contaminated?

3 Do you agree with the banning of DDT as an insecticide?

4 Do you agree that the sanitary landfill is the best way to dispose of solid wastes? Can you think of a practical alternative to this method?

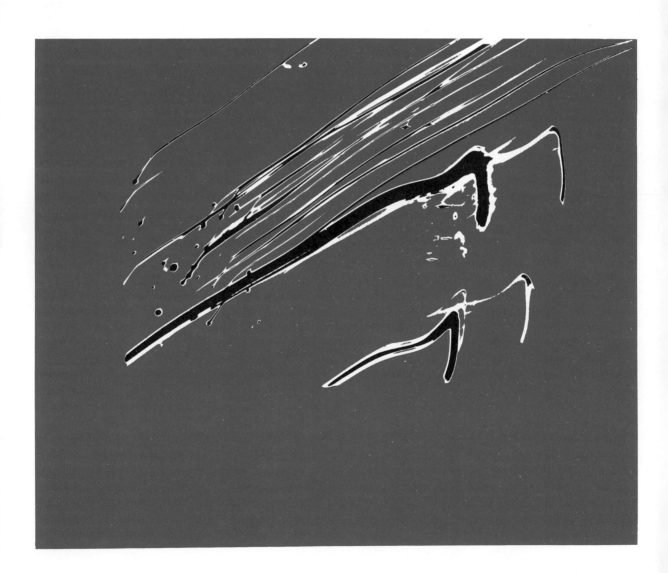

chemistry and the energy crisis

In recent years, it has become apparent that the United States is rapidly consuming its energy reserves. Although this situation had been predicted, it was not obvious to most Americans until the winter of 1972–73, when fuel-oil shortages appeared in the Midwest, and the summer of 1973, when many service stations ran out of gasoline. Since that time, the situation has become alternately better and worse, and periodic predictions of doom for the United States and world economies have shaken people's confidence. At the same time, there have been counterclaims that the so-called energy crisis is merely a plot on the part of energy producers to raise prices.

energy carrier
solar cell
semiconductor
solar farm

the importance of chemistry
in the energy supply

As the economy of the United States has grown, the need for energy has grown also. Electricity, for example, is used extensively for producing such large-volume industrial substances as acetylene, aluminum, chlorine, and magnesium, in addition to its obvious uses for heating, air conditioning, and lighting. Petroleum is consumed in large quantities for transportation and heating. Natural gas is used for both industrial and residential purposes.

It is logical to wonder what the energy source is for these operations. The ultimate answer is that the energy comes from the nuclear reactions that take place primarily on the sun. The light energy emitted by the sun is converted to chemical energy by the process of photosynthesis in green plants. Over the years, part of the chemical energy in plant and animal tissues has been stored beneath the surface of the earth in the form of coal and petroleum. Thus coal and petroleum are organic materials containing large amounts of energy that may be released by combustion to carbon dioxide and water.

$$petroleum + O_2 \rightarrow CO_2 + H_2O + energy$$

The function of coal and petroleum in society is that of an **energy carrier** No one really is interested in owning a gallon of gasoline; he is interested only in the energy that can be obtained by burning the gasoline. By burning a chemical energy carrier, the energy is released as heat energy that then may be converted into mechanical or electrical energy. Common chemical energy carriers are gasoline, diesel fuel, heating oil, coal, coke, and natural gas.

the case for an energy crisis

Since World War II, petroleum probably has been the major source of our energy. Petroleum and petroleum derivatives are used to generate electricity, heat homes, power automobiles, and operate factories. Other sources of energy, such as nuclear power and coal, have been only marginally competitive, owing to the low cost of petroleum. Consequently, worldwide petroleum consumption has reached a tremendous rate. In 1972, for example, consumption was estimated at 49.7 million barrels per day (a barrel is 42 gallons), with proved reserves of about 670 billion barrels. For both political and technical reasons, production of petroleum since that time has not always been able to meet the demand.

The results of insufficient production are quite apparent. Gasoline, heating oil, and other petroleum-based fuels have often been in short supply. Prices have risen sharply, and various conservation measures have been attempted. Other petroleum-based products such as plastics and synthetic fibers have periodically been hard to obtain also. Brownouts and temporary blackouts have occurred or have been threatened in various parts of the country.

Although the situation in the United States has been uncomfortable, the situation in other parts of the world has been even worse at times. During the Arab oil boycott in the winter of 1973–74, Britain had to adopt a temporary three-day workweek to cope with coal and petroleum shortages. At that time, some European countries banned Sunday driving, and the economy of Japan was hurt significantly.

The case for the energy crisis could be made even stronger. It is more important here, however, to consider the alternatives that have been, or might be, developed to solve the problem. It must be recognized that resolution of the problem will be slow and that no single solution can be expected to solve it entirely. For example, it will take time for automobile manufacturers to convert from making large cars to making smaller cars with better fuel economy. In the meantime, many larger cars still will be on the road. As another example, new homes can be built with better insulation, but it will take time to insulate or replace many older homes.

There are several obvious sources of alternate fuels that might be considered in place of petroleum. A few of these are nuclear power, coal, and solar energy. Nuclear power plants have been developed already; they were discussed in Chapter 5. Extensive research will be required before they will be both economical and safe to operate in large numbers, however.

coal

Coal already is being used extensively for the production of electricity in many parts of the nation. It is one resource for which the supply appears to be plentiful at the present. Authoritative estimates are that the United States has a 500-year supply at the present rate of consumption. Unfortunately, the problems of mining have not been thoroughly worked out yet. The safest and most economical method of mining coal is strip-mining, although underground mining is still practical.

In addition to direct conversion to electrical power, there are other ways that coal can be used to provide energy. Probably the method that will be used most extensively in the near future will be conversion of coal to natural gas, or methane. The fundamentals of this operation, called *coal gasification*, are relatively simple. Coal is heated with steam at high temperatures to generate carbon monoxide, carbon dioxide, and methane.

$$2 \text{ C} + 2 \text{ H}_2\text{O} \rightarrow \text{CO}_2 + \text{CH}_4$$

$$3 \text{ C} + 2 \text{ H}_2\text{O} \rightarrow 2 \text{ CO} + \text{CH}_4$$

The chemistry of the process has been known for years, but the process was not economical so long as plentiful supplies of natural gas were available. The pollution problems associated with coal gasification were discussed in Chapter 17. Several pilot plants for coal gasification are already in operation, and plans for full-scale plants are well developed.

The reason for the sudden interest in synthetic natural gas from coal is that existing supplies of natural gas are being used faster than new supplies are being located. Consequently, some natural gas is being imported at a cost that is about double that of domestic supplies. At present, synthetic natural gas can be manufactured at a cost that is about three times the cost of domestic natural gas.

When natural gas supplies were plentiful, there was no incentive to develop synthetic natural gas, because of the price differential. Now that natural gas supplies are insufficient to meet the needs of the population, alternate sources must be found, even if they are more expensive. The net result will be an increase in the cost of natural gas. The advent of synthetic processes probably will place an upper limit on the cost increase at something like triple the prices of the early 1970s. Further research should permit production of synthetic natural gas at lower cost.

solar energy

Energy from the sun is abundant. Enough energy reaches the earth daily to more than meet the energy needs of the entire world. Unfortunately, solar energy is so dispersed that no economical way to use the energy directly has been found. (Farming is actually an indirect way to harvest solar energy, but usually is not thought of in that manner.) Several proposals have been made, however, which are being explored. These include use of solar cells and solar farms.

It has been possible for many years to convert sunlight directly into electricity using **solar cells**. Basically, a solar cell consists of a solid electrode, attached to the back of a semiconducting material, and a semitransparent metal-film electrode on the front of the **semiconductor**. (A semiconductor is a material that allows an electrical current to flow when the semiconductor is exposed to sunlight.) In a typical solar cell, cadmium sulfide crystals serve as the semiconductor. A silver electrode is connected to one side of the cadmium sulfide and a thin layer of indium is placed on the opposite side. Light striking the cadmium sulfide through the indium electrode releases electrons from the cadmium sulfide; the electrons then flow through the external circuit and back to the silver electrode. As long as light strikes the cadmium sulfide, electricity will flow. (See Figure 19-1.)

A number of materials have been used successfully in solar cells. Silicon and selenium, for example, have functioned commercially as semiconductors. Experimental cells have employed such diverse materials as copper oxide and anthracene.

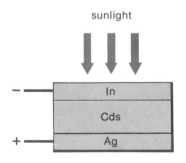

Figure 19-1. A solar cell

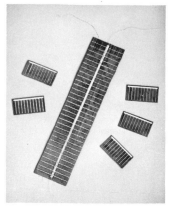

Solar cells (*above*) directly convert sunlight into electricity. They have been used extensively in space exploration and are being investigated as power sources for home and industry. (*Courtesy Tyco Laboratories, Inc., Waltham, Massachusetts*)

Artist's depiction (*above left*) of a solar farm, as conceived by A. B. Meinel. (*Artist: Dan Cowen, Tucson; photograph by George Kew*)

Solar One (*left*), an experimental project at the University of Delaware, is the first house capable of converting the sun's energy into heat and electricity. (*Courtesy Institute of Energy Conversion, University of Delaware, Newark, Delaware*)

Solar cells have been used to recharge the batteries used for space travel and the Skylab earth satellite. Peter Glaser, of the Arthur D. Little Company, has suggested that a huge satellite might be placed in orbit around the earth with a panel of solar cells perhaps five miles in diameter. The power collected by such a satellite could be beamed to the earth via microwaves. This proposal has the advantage that the satellite would be exposed to the sun almost the entire day and would not be subject to interference from clouds.

Unfortunately, solar cells are expensive, and many are needed to generate large quantities of electricity. Thus extensive use of solar cells has not been achieved. Further research, however, should permit the manufacture of more efficient solar cells at lower cost.

An alternate proposal for collecting solar energy comes from A. B. Meinel at the University of Arizona, who believes that temperatures as high as 550°C might be achieved in special tubes designed to absorb, but not re-emit, heat energy. A large number of these tubes located in sunny, arid parts of the world such as Yuma, Arizona, could run ordinary steam generators (and incidentally desalinate water at the same time). These large solar collection stations are referred to as **solar farms**

In any solar-energy collection process, energy storage becomes a problem that must be considered, as nighttime and cloudy days make the system inoperative. At the same time, however, the need for energy is continuous. The solar farm plan involves an energy-storage system as well as an energy-collection system. In the energy-storage system, heat energy is used to melt a salt mixture located in insulated tanks. When the energy is needed, water is passed through pipes located in the tanks. The hot salt vaporizes the water, and the resulting steam is used to generate electricity. It is believed that a nearly constant temperature of about 550°C could be achieved by choosing a salt mixture that melts at that temperature.

Prior to its devastating accident in 1937, the Hindenburg had successfully completed 54 flights, 36 of them across the Atlantic Ocean. While docking at Lakehurst, New Jersey, on May 6, 1937, the Hindenburg caught on fire and crashed, killing 36 persons. It is believed that an electrical discharge in the atmosphere near a hydrogen leak on the airship ignited the hydrogen and caused the ship to burn. Fifteen years before this, following the crash and burning of the U.S. airship Roma that killed 34 persons, the United States had ruled that only nonflammable helium could thereafter be used in any of its airships.

---

hydrogen as an energy carrier

No matter what energy source is used, the problem of distribution must be considered. Presently, energy is distributed through power cables as electrical energy or through pipelines and trucks as chemical energy (natural gas, propane, butane, gasoline, and so on). Hydrogen has been proposed as an alternative energy carrier as the natural sources of chemical fuels are depleted.

Hydrogen is generated easily by the electrolysis of water, using a small quantity of sulfuric acid as a catalyst.

$$2 \text{ H}_2\text{O} \xrightarrow[e^-]{\text{H}_2\text{SO}_4} 2 \text{ H}_2 + \text{O}_2$$

A by-product of the process is oxygen, which has many industrial and medicinal uses. The energy required for this reaction is provided by electricity and is chemically stored in the products, hydrogen and oxygen. Combustion of the hydrogen with oxygen releases this stored energy as heat. Since the combustion product is water, no pollution problem is foreseen.

Hydrogen could be generated locally for cooking and heating, or it could be generated at the source of electrical power and transported by pipeline to the point of consumption. Electricity can be generated directly from hydrogen using fuel cells (discussed in Chapter 20) or indirectly by conventional steam generators. Even automobiles can be powered by hydrogen instead of gasoline. Figure 19-2 is a diagram showing the possible generation, transport, and use of hydrogen as an energy carrier.

Extensive use of hydrogen as a fuel has been hindered because of the explosive danger of hydrogen and air mixtures. Recent advances in storing and handling hydrogen have resulted from the space program, where large quantities of liquid hydrogen have been

---

Methanol, or wood alcohol (CH_3OH), also has been suggested as an energy carrier. It has the advantages of being inexpensive, easy to manufacture, and clean burning. Although presently made only from natural gas or petroleum, it may someday be made from coal. Economic projections suggest that it could be made from coal at a price competitive with gasoline for automobile use. Up to 10% methanol may be added to gasoline without any modification of the engine. Complete conversion to methanol would require some changes in engines and fuel tanks, since it takes about two gallons of methanol to produce the same amount of energy as one gallon of gasoline.

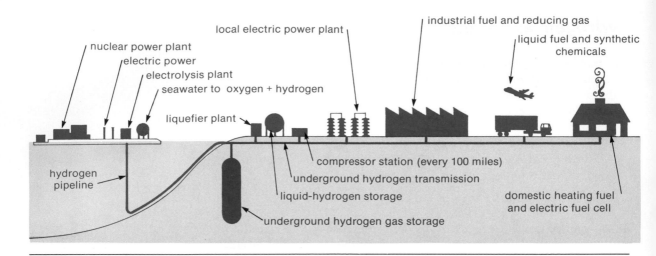

nuclear power plant
electric power
electrolysis plant
seawater to oxygen + hydrogen
liquefier plant
local electric power plant
industrial fuel and reducing gas
liquid fuel and synthetic chemicals
hydrogen pipeline
compressor station (every 100 miles)
underground hydrogen transmission
liquid-hydrogen storage
underground hydrogen gas storage
domestic heating fuel and electric fuel cell

Figure 19-2. Proposed scheme for generating, storing, and using hydrogen as an energy carrier. Hydrogen, generated from sea water, can be stored underground for later use or used directly in the generation of energy for household and industrial purposes. (Redrawn from a diagram by Nicholas Fasciano, *Fortune*, Nov. 1972)

required as rocket fuel. As a consequence, many people now claim that hydrogen is as safe to handle as existing chemical energy carriers.

No matter what solutions are found for the immediate energy crisis, long-range changes in the world's energy supplies are inevitable. Chemical energy carriers will be needed as part of the new energy supply, just as they are needed presently. Exhaustion of natural energy sources will increase the price of energy somewhat because of the need to utilize new sources. It can be hoped that new scientific developments will keep price increases below those currently projected for alternate energy sources.

references

Faltermayer, Edmund, *The Energy "Joyride" Is Over*, Fortune, **86** (3), 99 (1972).

Gilmore, C. P., *On the Way! Plentiful Energy from the Sun*, Popular Science, **201** (6), 86 (1972).

Lessing, Lawrence, *The Coming Hydrogen Economy*, Fortune, **86** (5), 138 (1972).

Lessing, Lawrence, *Capturing Clean Gas and Oil from Coal*, Fortune, **88** (5), 129 (1973).

Krieger, James H., *Energy: The Squeeze Begins*, Chemical and Engineering News, **50** (46), 20 (1972).

Saudi Arabia—A Nation We'd Better Get To Know, Forbes, **111** (4), 28 (1973).

study questions

1 Name five energy carriers.
2 Do you believe that there is an energy crisis?
3 Discuss three possible new energy sources.
4 What is the ultimate source of energy used on the earth?
5 What are some advantages of coal gasification as a source of fuel?
6 What are some disadvantages of coal gasification as a source of fuel?
7 How is a solar cell constructed?
8 Why does a solar cell produce electricity?
9 List several ways in which solar energy might be collected.
10 Describe a solar farm.
11 How might hydrogen be used as an energy carrier?
12 If hydrogen were used as an energy carrier, how would it be generated?

special problems

1 It has been suggested that selected plant materials might be grown on large farms specifically for their fuel values. Is this a particularly efficient way to collect solar energy?
2 The combustion of approximately 2 gallons of methanol produce the same amount of energy as 1 gallon of gasoline. In the future, it may be possible to make methanol from coal at a price that is competitive with gasoline. How large would the gas tank have to be if methanol is the fuel for a car that now gets 15 miles per gallon of gasoline and can travel 300 miles on one tank of gasoline?
3 Is it practical to have an automobile fuel tank of the size determined in Problem 2?
4 Trace the path of energy from the sun to the heat from an electrical heater.
5 Discuss other possible sources of energy that have not been mentioned in this chapter.

chapter **20**

chemistry in space

Man has always been fascinated by phenomena and objects in outer space. Observations of the regular movements of stars enabled the Egyptians to devise an accurate calendar as early as 4000 B.C. Shortly before the time of Christ, Greek astronomers had concluded that the earth was spherical and had actually calculated its size. In spite of this early work, until after Columbus discovered America most people still believed that the earth was flat. Invention of the telescope by Galileo and calculations by Nicolas Copernicus, which showed that the sun was the center of the solar system and that the earth rotated around the sun, were followed by rapid advances in man's understanding of outer space.

Until relatively recently, all man's understanding of outer space was the product of observations made from the earth. Now, advances in rocket technology (primarily since World War II) permit investigations in outer space itself. The principal agency in the United States that is responsible for the space program is NASA.

watt-hour
fuel cell
life-support system
activated carbon

nasa

The National Aeronautics and Space Administration, commonly called NASA, was established by the National Aeronautics and Space Act of 1958. In this act, the Congress of the United States directed NASA to work toward the expansion of human knowledge of the atmosphere and space. NASA was further directed to disseminate the information gained as widely as possible.

Since its founding, NASA has sent unmanned space probes to Venus, Mars, Jupiter, and other planets. It has sent men to the moon and obtained samples of moon rocks and soil. Investigations into the age and composition of the moon, the atmosphere of Venus, and the atmosphere and surface of Mars are just a few of the important

Apollo Little Joe II launch vehicle lifts from its launching pad. Principal objective of the test was to prove the performance of the launch escape subsystem. (*Courtesy NASA*)

Earth, as seen from Apollo 17 spacecraft. (*Courtesy NASA*)

studies that have been conducted by NASA. All these activities have added greatly to our knowledge of the other planets in our solar system.

The direct and indirect benefits of the NASA program have been impressive. Earth satellites have made direct television communication around the world a reality. Weather forecasting has become significantly more reliable, now that daily satellite photographs of weather formations are available. This increased reliability has been of tremendous value to people in areas subject to hurricanes and other severe weather conditions. The complexity of space navigation and communications has led to tremendous advances in computer and radio technology. Further benefits, even greater than those already achieved, will undoubtedly become obvious over the next several years.

Since ancient times, man has had a desire to travel to outer space. The obstacles that had to be overcome in order to accomplish this objective were tremendous. Huge rocket ships had to be de-

Tyros II weather satellite being tested. Solar cells on the satellite's top and sides are checked with the aid of lights mounted on the test rack. (*Courtesy NASA*)

Although Hurricane Becky, shown in a weather-satellite photo taken in August, 1974, may present a hazard to ships in the area, it does not appear to pose an immediate danger to the eastern seacoast of the United States. (*Courtesy National Oceanic and Atmospheric Administration*)

Picture taken of the earth from outer space shows clearly the cloud cover, including a hurricane off the Texas coast and a tropical storm developing off the African coast. (*Courtesy National Oceanic and Atmospheric Administration*)

Beneficial uses of man-made satellites undoubtedly will increase. Currently, these satellites are used in air pollution studies, collecting navigational information, measuring factors affecting crops (such as the fungus blight in the Midwest a few years ago), and locating promising sites of natural resources.

signed that could take a sizable amount of equipment beyond the earth's gravity. Heat-resistant nose cones had to be developed for reentry of rockets into the earth's atmosphere. Long-range, reliable power supplies were required, as well as life-support systems for manned flights.

As man ventures further into space, existing problems will be magnified, and others will develop. As in the past, many problems will be chemical and will require chemical solutions. Rocket fuels are chemicals. Energy storage usually is accomplished via batteries or fuel cells that are chemical entities. Life-support systems are largely

chemical systems designed to provide the proper chemicals (oxygen and food) to the astronauts and remove such by-products as carbon dioxide. Some of the existing and proposed chemical solutions of these problems will be discussed in this chapter.

Comparison of two Skylab rockets. *Left*: Skylab II space vehicle. *Right*: Skylab I space vehicle. Note relative sizes of Skylab II rocket and ground vehicles. (*Courtesy NASA*)

A feeling for the magnitude of the problems of just a short spaceflight may be obtained by considering some of the statistics related to the Apollo 11 mission (first lunar landing). The Saturn V rocket system was 363 feet high (as high as a 36-story building). The command module alone contained over 2 million parts. Of the nearly 6.5 million pounds that the system weighed, over 6.0 million pounds represented fuel, consisting of kerosene and liquid oxygen for the first stage and liquid hydrogen and liquid oxygen for the second and third stages. Designing a system that would hold the liquid hydrogen (at $-253°C$) and liquid oxygen (at $-183°C$) without becoming brittle and cracking at the low temperature and that also would withstand the high temperatures in the rocket engine itself was a major achievement.

rocket engines and rocket propellants

Science-fiction writers dream about interplanetary and even interstellar travel by means of "flying saucers" powered by antigravity engines or other means of propulsion. Meanwhile, all space exploration that has been conducted to date and all that is contemplated for the foreseeable future is based on the simple rocket engine using a chemical fuel.

The rocket engine provides the accelerating thrust necessary to move the rocket. The thrust is achieved by expelling small particles of propellant from the engine at high speed. The thrust is related to the amount of propellant expelled in a given time and the velocity of the propellant particles. Figure 20-1 illustrates a rocket engine.

The energy for expelling the propellants comes from a combustion reaction. The average velocity of the propellant molecules (and thus the thrust) is increased by increasing the combustion temperature and decreasing the average size of the combustion products. In other words, the ideal fuel will evolve a large amount of energy and produce small particles.

There are basically two types of rocket-propellant systems currently in use, solid and liquid. To function, each type requires a fuel and an oxidizing agent. In some solid propellants, such as nitrocellulose and nitroglycerine, both functions may be built into the same molecule. In other solid propellants, the oxidizing agent is a separate compound that is mixed with the fuel. The fuel is frequently a polymer that helps to bind the entire mass together.

Rockets powered by solid propellants have the advantage that they are simple to design, since no special fuel tanks or pumps are required. They have the disadvantage that they cannot be shut down once they are fired, and they are subject to shock and chemical decomposition during storage. The Polaris missile and the Nike booster are examples of solid-propellant rockets.

Normally, liquid propellants are used in applications where it is desirable to control the time that the engine is in operation. Directional changes or other maneuvers usually are made by using liquid-propellant rockets. Since the oxidant and the oxidizing agent are separated, this type of rocket propellant tends to be safer to handle. Liquid oxygen (LOX) is a common oxidant that is used with alcohol (as in the German V-2), with a special form of kerosene called RP-1 (as in the first stage of the Saturn V-Apollo booster), or with hydrogen (as in the third stage of the Saturn V-Apollo booster). Liquid oxygen has a disadvantage, because it must be stored at −183°C, or under high pressure. Low temperature necessitates insulation, and high pressure requires reinforced containers. Both add weight to the

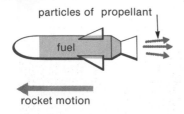

particles of propellant

fuel

rocket motion

Figure 20-1. A rocket engine

rocket, reducing the payload. Other oxidants, however, frequently weigh more for a given oxidizing capability, also adding weight and reducing the payload. As with most systems associated with space flight, the final choice of fuel and oxidant results from a compromise between the advantages and disadvantages in a given system.

electrochemical cells

Electrical power is a necessity in space exploration at both the manned and unmanned levels. Chemical cells and batteries have been employed in all space vehicles. The fundamental principles of chemical cells were discussed in Chapter 10.

For space travel, a battery with a high power density is necessary. In other words, the battery must deliver a large amount of electricity and weigh as little as possible. To cut down further on weight, the battery should be rechargeable from some external energy source such as the sun. The familiar lead storage battery is rechargeable, but much too heavy for space application.

Space scientists have developed a lighter battery that uses silver oxide and zinc electrodes and a potassium hydroxide electrolyte. This cell has a power density of 40 watt-hours per pound. In other words, a one-pound battery can be expected to produce 40 watt-hours of electricity. A watt-hour is one one-thousandth of the kilowatt hour used by power companies to measure power for home or industry. Unfortunately, the silver oxide–zinc battery can be recharged only about 80–100 times before further recharging is ineffective. This battery is adequate for short trips, such as to the moon, but is not suitable for long-range trips or satellites that must function for a year or more. Nickel-cadmium batteries, similar to those in cordless appliances, may be used in long-range situations. They have a lifetime of 1000 to 11,000 recharging cycles, but a power density of only 7–10 watt-hours per pound.

Since space exploration is expected to involve extremes of temperature, a number of cells have been designed that can be used under these conditions. Ordinary cells using aqueous electrolytes would boil or freeze under some possible conditions. Cells involving molten-salt electrolytes have been developed for high-temperature applications. Liquid ammonia has been used for low-temperature cells.

Another type of cell, the fuel cell, also has been employed in space travel. The fuel cell consists of two porous electrodes connected by an electrolyte. The oxidizing agent (usually oxygen) and the

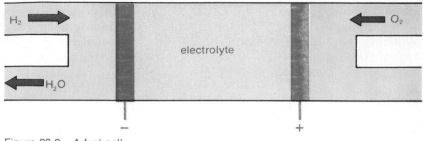

Figure 20-2. A fuel cell

reducing agent (usually hydrogen) are introduced as gases at the two electrodes. A fuel cell is shown in Figure 20-2.

The chemistry for the fuel cell is identical to that of an ordinary electrochemical cell.

Positive electrode $\quad\quad 2\,H^+ + \dfrac{1}{2}\,O_2 + 2\,e^- \rightarrow H_2O$

Negative electrode $\quad\quad\quad\quad\quad\quad H_2 \rightarrow 2\,H^+ + 2\,e^-$

Net reaction $\quad\quad\quad\quad\quad\quad H_2 + \dfrac{1}{2}\,O_2 \rightarrow H_2O$

Advantages of the fuel cell are that it does not run down and that it uses the same fuel as some rocket engines. In addition, the by-product, water, can be applied to other purposes.

life-support systems

When man travels in space, it is necessary for him to carry along the appropriate necessities to support life. These are primarily food, oxygen, and water. It is also necessary to carry along waste-disposal equipment. The supplies and equipment that comprise the life-support system must be compact, simple, and reliable. For short trips, such as the moon landings that have been carried out, all these supplies have been stored aboard the spacecraft.

carbon dioxide

Carbon dioxide removal has been necessary even for short trips. An average man produces about a kilogram of carbon dioxide per day. The maximum safe concentration of carbon dioxide in the atmo-

Astronaut walking on the moon depends totally on the life-support system carried on his back. (*Courtesy NASA*)

sphere is about 3.7%. Higher concentrations cause headache, dizziness, and impaired thinking. Sufficiently high concentrations or long exposure lead to death.

For the short flights to the moon and back, carbon dioxide removal has been accomplished entirely by absorption with lithium hydroxide.

$$2\ LiOH + CO_2 \rightarrow Li_2CO_3 + H_2O$$

Lithium hydroxide was chosen rather than another alkali metal hydroxide because of its low molecular weight. As shown on page 140, only about 1.1 kilograms of lithium hydroxide are required to absorb 1 kilogram of carbon dioxide. By the same procedure, it could be shown that 1.8 kilograms of sodium hydroxide or 2.5 kilograms of potassium hydroxide would be required to do the same job. In space travel, where weight is more important than initial cost (sodium hydroxide would be the least expensive), lithium hydroxide is the obvious choice.

For longer flights, the amount of lithium hydroxide required becomes prohibitive. For example, even a relatively short, 30-day trip would require 330 kilograms or a little over one-third of a ton of lithium hydroxide to absorb all the carbon dioxide produced by one man. Flights to Mars or other planets that might take a year or more are impractical until we solve the problem of carbon dioxide removal.

Several approaches to the problem are currently under investigation. One system that looks promising involves use of silver oxide. Silver oxide will combine with carbon dioxide to produce silver carbonate, much like the lithium hydroxide system that has been discussed already.

$$Ag_2O + CO_2 \rightarrow Ag_2CO_3$$

Silver oxide, however, is regenerated easily by heating the silver carbonate at 180°C for four hours. The carbon dioxide produced by this regeneration is concentrated and may be vented out of the spacecraft, or it may be used to regenerate oxygen.

oxygen

For short flights, oxygen has been carried in the spacecraft as compressed gas or liquid. For longer flights, the weight of the required oxygen supplies will be excessive. In this situation, some form of oxygen regeneration will be required.

Several methods of regenerating oxygen have been studied. One procedure that looks reasonable involves the following reactions.

$$CO_2 + 4\ H_2 \xrightarrow[200-260°C]{Ni} CH_4 + 2\ H_2O$$

$$CH_4 \rightarrow C + 2\ H_2$$

$$2\ H_2O \xrightarrow{e^-} 2\ H_2 + O_2$$

$$CO_2 \rightarrow C + O_2$$

The net effect of these reactions, if the hydrogen is recycled, is shown at the left. Water also could be converted to oxygen and hydrogen in this process.

The cabin atmosphere used in the Apollo 11 mission, during which the first lunar landing occurred, consisted of 60% oxygen and 40% nitrogen. Pure oxygen had been used in some earlier space flights, but a tragic fire that killed three astronauts during a practice session showed this procedure to be unsafe.

trace contaminants

Numerous sources of trace contaminants exist in the closed atmosphere of a spacecraft. Combustion by-products, rocket propellants, lubricants, and metabolic wastes are just a few of the hazardous contaminants. The most effective method of removing these materials from the spacecraft's atmosphere has been adsorption on the surface of activated carbon. Activated carbon is a finely powdered form of carbon that has a large total surface area for interaction with contaminants. Many substances are readily adsorbed onto this surface. The same principle operates in most gas masks and some cigarette filters.

molecules in space

Until relatively recently, it was thought that the space between stars and planets was largely unoccupied. More recent discoveries indicate that this is not true. In the past few years, many molecules have been identified in interstellar space. Some of the more important are listed below.

Carbon monoxide	CO
Ammonia	NH_3
Hydroxyl radical	OH
Water	H_2O
Hydrogen cyanide	HCN
Formaldehyde	$HCHO$
Formic acid	$HCOOH$
Methanol	CH_3OH
Cyanoacetylene	$HC\equiv CCN$

The distribution of these molecules in space has led to speculation about their role in the formation of stars and the evolution of life. Formaldehyde, for example, appears to be widely dispersed throughout the Milky Way galaxy. Water and ammonia, however, appear to be concentrated in regions about the size of our solar sys-

Nebula in *Scutum Sobieski.* Nebulae are swirling or rotating masses of gas and dust particles that astronomers believe could condense to form stars. (*Photograph reproduced by permission of the Hale Observatories*)

tem. If the molecules in these regions were to coalesce, owing to gravity, into smaller, more dense regions, nuclear reactions might occur. It has been suggested that these regions represent stars in the process of being formed, since nuclear reactions have been associated with the energy-producing process in stars.

Theories on the origin of life on earth have assumed that the early atmosphere consisted of such molecules as methane, formaldehyde, and ammonia. Experiments with electrical discharges in flasks containing these and other materials thought to exist in the early atmosphere have shown that such important biological molecules as amino acids could have been formed from such an atmosphere. One important flaw in these theories has been that the origin of these molecules was hard to explain. It now appears that some of these molecules already exist in space. (Methane and amino acids have not yet been identified in space, but many scientists believe that they soon will be found.) Thus it is possible that life actually originated in space. At any rate, the presence of these "pre-life" molecules in space makes the origin of life on earth easier to rationalize.

chemistry of the moon

Since the first manned landing on the moon, our understanding of the origin of the solar system has advanced significantly. The oldest rocks found on earth are about 3.5 billion years old. Some of the lunar rocks, however, appear to be as much as 4.5 billion years old. Since there is essentially no atmosphere on the moon, the surface of the

Picture of Charles M. Duke, Jr., collecting lunar samples, taken by John W. Young. The Lunar Roving Vehicle is seen in the left background. (*Courtesy NASA*)

moon is not subject to the normal weathering observed on the earth. Since the solar system also is believed to be about 4.5 billion years old, scientists finally have been able to examine rocks that they believe were formed at the same time as the earth.

Analysis of the moon rocks indicates that, compared with the natural occurrences of the elements on earth, on the moon there is a higher abundance of the higher-melting elements, such as titanium, and a lower abundance of the lower-melting elements, such as sodium. The current explanation suggests that when the material was molten, the lower-melting materials were more volatile and escaped into space. It must be recognized, however, that the data available so far represent samples collected at only a few locations on the moon and may not be typical of the entire moon. Confirmation of these suggestions and hypotheses must wait for analyses of further samples of moon rocks.

future endeavors

Space exploration has really only begun. The potential for future accomplishments is unlimited. One can visualize interplanetary and instellar travel, colonization of distant planets, and other endeavors presently found only in science-fiction stories.

To make these dreams a reality, further developments are necessary. For example, new modes of space travel, perhaps even at speeds greater than the velocity of light (now regarded as theoretically impossible), are required. Improved life-support systems, protective devices, construction materials, communication systems, and other equipment also must be developed. The potential to solve many of these problems now exists, and the realization of these objectives can be expected in the near future.

references

Bonn, Bertram, *Chemistry among the Stars*, Men and Molecules Program 461, American Chemical Society, Washington.

Lawrence, Richard M., *Space Resources for Teachers—Chemistry*, National Aeronautics and Space Administration, Washington, 1971.

Buhl, D., and L. Snyder, *Molecules in Space*, Men and Molecules Program 512, American Chemical Society, Washington.

study questions

1 Name some tangible benefits of the United States space program.
2 What is the function of a rocket fuel?
3 What characteristics make a good rocket fuel?
4 What are the relative merits of solid and liquid rocket propellants?
5 What is the difference between a fuel cell and a voltaic cell?
6 What is the advantage in a rocket ship of a fuel cell over a voltaic cell?
7 In what unit is electrical energy usually measured?
8 How is the carbon dioxide concentration in a rocket-ship cabin controlled? Why is a lithium salt used, as opposed to a sodium or potassium salt, for example?
9 Describe a possible process by which oxygen regeneration might be accomplished for use in space travel. Why is there interest in developing processes such as this?
10 What interest is there in studying molecules found in space?
11 How has the discovery in space of molecules such as ammonia and formaldehyde affected our concept of the origin of life?
12 What are some scientific results of the United States space program?

special problems

1 Discuss the similarities between a charcoal cigarette filter and the method used for trace-contaminant removal in a spacecraft.
2 What kind of projects should NASA undertake next?
3 Should the space program be expanded, contracted, or continued at the present level?

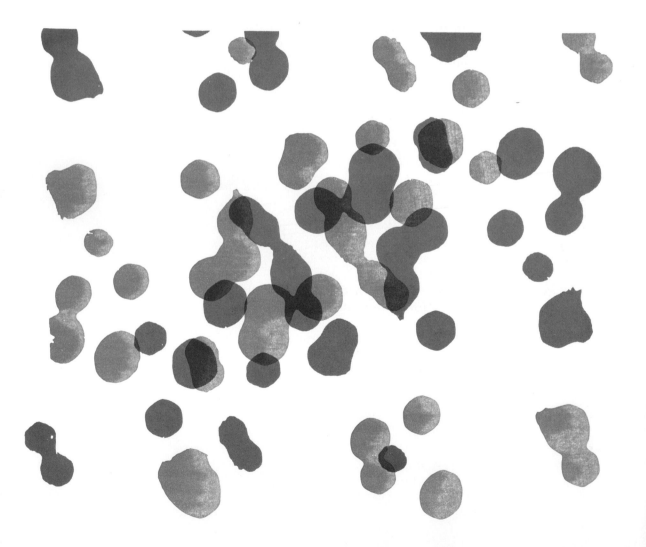

epilog

It is appropriate, at the end of a course of study, to reflect on what has been gained as a result of that study. As the authors, we should like to comment on our thoughts concerning some of the long-term benefits that we hope you have derived from your efforts. We realize that ten years from now most of you will not remember how to balance chemical equations, but we hope that you will remember that chemistry is a quantitative science and that you could draw on that knowledge if the need should arise. We know that you will not long remember how to write the electronic structure of an element, but we hope that you will find satisfaction in retaining at least a qualitative understanding of the structure of the atom.

Probably more important, however, we hope that you will see the proper place of science in society. Science is not the solution to all man's problems, as many seemed to think during the early 1960s. At the same time, it is not the cause of all man's problems, as many seemed to think in the early 1970s. Science is merely an organized search for knowledge, much of which may be applied to technology. Some applications will be beneficial and some not. It is the responsibility of all people, scientists and nonscientists alike, to weigh the advantages and the disadvantages of a technological advance before applying it to the service of mankind. Furthermore, continuing appraisal is needed to ensure that previously unrecognized problems are detected before they become major ones.

Some of today's college students will be business or political leaders in just a few years. They will be required to function responsibly in a society that probably will have become even more technologically oriented than today's. Many problems, such as those involved with space flight, pollution control, metric conversion, and energy, will have to be solved. Proper allocation of available resources between social welfare and research probably will continue to be a concern at both national and local levels.

We hope that your present effort in the study of chemistry will be of value as you face the problems of the future. As a concerned citizen it will be your responsibility to see that intelligent decisions are made regarding these problems. It is our belief that a person with scientific training is better equipped to evaluate problems and potential solutions of a technological nature than a person who lacks this training.

In conclusion, then, we hope that your time spent in studying chemistry has been of some enjoyment and value, both immediate and long term.

index